D1214986

Advances in MATHEMATICAL ECONOMICS

Aims and Scope. The project is to publish *Advances in Mathematical Economics* once a year under the auspices of the Research Center of Mathematical Economics. It is designed to bring together those mathematicians who are seriously interested in obtaining new challenging stimuli from economic theories and those economists who are seeking effective mathematical tools for their research.

The scope of *Advances in Mathematical Economics* includes, but is not limited to, the following fields:

- Economic theories in various fields based on rigorous mathematical reasoning.
- Mathematical methods (e.g., analysis, algebra, geometry, probability) motivated by economic theories.
- Mathematical results of potential relevance to economic theory.
- Historical study of mathematical economics.

Authors are asked to develop their original results as fully as possible and also to give a clear-cut expository overview of the problem under discussion. Consequently, we will also invite articles which might be considered too long for publication in journals.

Springer
Tokyo
Berlin
Heidelberg
New York
Barcelona
Hong Kong
London
Milan
Paris
Singapore

S. Kusuoka, T. Maruyama (Eds.)

Advances in Mathematical Economics

Volume 1

Shigeo Kusuoka
Professor
Graduate School of Mathematical Sciences
University of Tokyo
3-8-1 Komaba, Meguro-ku
Tokyo, 153-0041 Japan

Toru Maruyama
Professor
Department of Economics
Keio University
2-15-45 Mita, Minato-ku
Tokyo, 108-8345 Japan

ISBN 4-431-70251-2 Springer-Verlag Tokyo Berlin Heidelberg New York

Printed on acid-free paper

Printed in Japan

Photocomposed copy prepared from the authors' electronic files.
Printed and bound by Hirakawa Kogyosha, Japan.
SPIN: 10707913

Editors' Preface

Advances in Mathematical Economics is a publication of the Research Center of Mathematical Economics[†], which was founded in 1997 as an international scientific association that aims to promote research activities in mathematical economics.

Our new publication was launched to realize our long-term goal of bringing together those mathematicians who are seriously interested in obtaining new challenging stimuli from economic theories and those economists who are seeking effective mathematical tools for their research.

The first issue of *Advances in Mathematical Economics* is essentially a collection of papers presented at the Symposium on Mathematical Analysis in Economic Theory, which was held at Keio University (Tokyo) during October 4–5, 1997, on the occasion of Professor Gérard Debreu's first visit to Japan.

The topics discussed in the symposium may be classified under the following four headings:

- Nonlinear Dynamical Systems and Economic Fluctuations.
- Stochastic Analysis and Mathematical Finance.
- Convex Analysis and Optimization Theory.
- Aspects of General Equilibrium Theory.

It is a great pleasure to publish this volume consisting of significant contributions by true leaders in the discipline. On behalf of the editorial board, we would like to express our cordial gratitude to all the participants for their warm and constructive contributions, and also to

All Nippon Airways Co., Ltd.
Japan Society for the Promotion of Science
Keio University

for their generous financial support.

September 16, 1998
Shigeo Kusuoka
Toru Maruyama

[†] Office: Toru Maruyama, Department of Economics, Keio Unversity, 2-15-45 Mita, Minato-ku, Tokyo 108-8345, Japan

Table of Contents

Note

Appendix

Adv. Math. Econ. 1, 1–2 (1999)

Advances in
MATHEMATICAL
ECONOMICS

On the use in economic theory of some central results of mathematical analysis

Gérard Debreu

Department of Economics, University of California, Berkeley, CA 94720-3880, USA
(e-mail: debreu@econ.berkeley.edu)

Received: March 26, 1998

JEL classification: D5

Mathematics Subject Classification (1991): 90A14

Four instances of fundamental theorems of economics established by means of central results in various branches of mathematical analysis are listed.

The first is a theorem giving conditions on the economy ensuring that, with any Pareto optimal state s, is associated a price-vector p with respect to which every agent is in equilibrium. Consider the set E, a subset of the commodity-space, whose generic element is an aggregate resource-endowment allowing the economy to achieve the state s. The actual aggregate endowment e of the economy is a boundary point of the set E, which is convex in two economically important cases. Then there is through e a hyperplane supporting E. This hyperplane has a normal p that is the desired price-vector.

The second is a proof of existence of a general equilibrium for a social system following J. Nash. For every i in $M = \{1, \cdots, m\}$, let A_i be the compact, convex, non-empty set of a priori possible actions of the ith agent. Once, $\forall i \in M$, the ith agent has chosen an action a_i in A_i, the state of the system is determined by $a = (a_1, \cdots a_m) \in A = X_{i=1}^{m} Ai$.The ith agent considers the actions chosen by the others $(a_1, \cdots a_{i-1}, a_{i+1}, \cdots, a_m) = a_{M\setminus i} \in A_{M\setminus i}$ as given. His optimal reaction is any point of $\mu_i(a_{M\setminus i})$, a subset of A_i. The correspondence μ from A to A defined by $a \to \mu(a) = \mu_1(a_{M\setminus 1}) \times \cdots \times \mu_m(a_{M\setminus m})$ is convex valued, and has a closed graph.

Every agent is in equilibrium if $\forall i, a_i \in \mu_i(a_{M\setminus i})$, i.e. if $a \in \mu(a)$. The set A is compact, convex, non-empty, and the correspondence μ is convex-valued, and has a closed graph. By Kakutani's theorem, μ has a fix point a^* that belongs to $\mu(a^*)$.

The third gives an explanation of the behavior of agents who, in an l-commodity economy, $\mathcal{E}$, consider prices as given. It is based on the game-theoretical concept of the core. Characterize the ith agent ($i \in M$) by his endowment e_i, a point of the closed positive orthant R^l_+, and his preference relation $>_i$, a binary relation on R^l_+ where "$z' >_i z$" is read as "z' is preferred by i to z". Let also $x = (x_1, \cdots, x_m)$ be called an allocation if $\forall i \in M, x_i \in R^l_+$ and $\sum_{i \in M} x_i = \sum_{i \in M} e_i$.

Coalition $I (\neq \phi)$ blocks allocation x if there is an allocation y such that

$$\sum_{i \in I} y_i = \sum_{i \in I} e_i \quad \text{and} \quad \forall i \in I, y_i >_i x_i,$$

$C(\mathcal{E})$ is the set of unblocked allocations.

The Walras set $W(\mathcal{E})$ is the set of allocations obtained by means of a price-vector that every agent considers as given.

Extension of these definitions from a finite set of agents to a measure space of agents $(A, \mathcal{A}, \nu)$, with a σ-field $\mathcal{A}$ of coalitions, and a positive measure ν such that $\nu(A) = 1$, is immediate.

Aumann has proved the following theorem.

If the measure space $(A, \mathcal{A}, \nu)$ is atomless, and the total endowment $\int_A e \, d\nu$ is strictly positive, then $W(\mathcal{E}) = C(\mathcal{E})$.

The proof can be based on a theorem of A.A. Lyapunov.

The fourth instance studies the problem of uniqueness of the economic equilibrium whose existence can be derived from the general theorem established in the second section. It proves, for example, that for differentiable economies with a finite number of agents and a finite number of commodities, the number of equilibria is finite except for a set of economics that is negligible in a strong sense. Let the economy be characterized by the initial endowments $e = (e_1, \cdots, e_m)$ of each agent, and let $\mathcal{E}$ be the $l \cdot m$-dimensional set of these m-lists.

The state of that economy can be described by the price-vector p (which may be restricted to belong to the positive part S of the unit sphere in the l-dimensional Euclidean space). By definition, p is an equilibrium price-vector if and only if the total excess demand $F(e, p)$ in R^l vanishes.

Let V be the equilibrium manifold, i.e., the set of points (e, p) of $\mathcal{E} \times S$ satisfying $F(e, p) = 0$, and let T be the projection from V to $\mathcal{E}$. A regular economy is defined as a regular value of T. According to Sard's theorem, the set Γ of critical economies has measure zero.

The set Γ is also easily proved to be closed.

Adv. Math. Econ. 1, 3–15 (1999)

Heterogenous probabilities in complete asset markets

Laurent Calvet[1], **Jean-Michel Grandmont**[2], **and Isabelle Lemaire**[3]

[1] Department of Economics, Harvard University, Littauer Center, Cambridge, MA02198, USA
[2] CREST and CNRS, 15 boulevard Gabriel Péri, 92245 MALAKOFF Cédex, France *
[3] CREST-INSEE, 15 boulevard Gabriel Péri, 92245 MALAKOFF Cédex, France

Received: April 13, 1998

JEL classification: C62, D50, G10.

Summary. We show in this paper how, in a model of assets exchange in complete competitive markets, heterogeneity of the agents' subjective probabilities generates aggregate expenditures for Arrow-Debreu securities that have the gross substitutability property, with the consequences that competitive equilibrium is unique, stable in any tatônnement process, and that the weak axiom of revealed preferences is satisfied in the aggregate. For this result, heterogeneity is required to be highest among people who have the largest risk aversion.

Key words: Heterogeneity, subjective probabilities, complete asset markets.

Introduction

It is well known that competitive general equilibrium models have little empirical content, in the sense that they can generate almost any aggregate behavior (up to Walras' law and absence of money illusion) when the distribution of individual characteristics (tastes, endowments) is arbitrary. It has been nevertheless shown recently (Grandmont (1992), Quah (1997)) that heterogeneity may in fact be helpful in that respect. In particular, if agents maximize utility of the form $u(x_1, \ldots, , x_n)$, where x_h is consumption of commodity h, one may consider a class of agents who have the same tastes up to some rescaling of the units of measurement of each commodity, i.e. who maximize a utility of the form $u(e^{\alpha_1} x_1, \ldots, e^{\alpha_n} x_n)$, where the parameters $\alpha = (\alpha_1, \ldots, \alpha_n)$ are arbitrary. Then if, say, the distribution in the class over

* Part of this research was done while the author was visiting CORE, Louvain la Neuve. Financial support from the Université Catholique de Louvain is gratefully acknowledged.

the vector α is sufficiently close to a uniform distribution with a large support, aggregate expenditures within this class become less sensitive to price variations, when income is price independent. This property, being additive, is preserved when one aggregates over different classes, in particular having fundamentally different tastes. In fact it generates, in competitive exchange economies where incomes, being the market value of the agents' initial endowments, depend on prices, aggregate excess demands that have the gross substitutability property, implying unicity and stability of equilibrium among things.

The purpose of the present paper is to investigate the possible benefits of heterogeneity in complete assets markets which, after all, are specific general equilibrium models, and are thus potentially subject to the kind of criticism we alluded to at the beginning of this introduction. We consider a standard one-period portfolio selection problem, where agents maximize expected utility $\sum_s \pi_s u(y_s)$, in which y_s is income in state s and π_s the subjective probability attached to that state. We focus on one essential source of heterogeneity among agents, namely heterogeneity of the subjective probabilities π_s, for two reasons. The first one is that the rescaling procedure mentioned above is no longer available in this context. For it would lead to the maximization of a state dependent utility of the form $\sum_s \pi_s u(e^{-\alpha_s} y_s)$, which would not be acceptable in models of finance (we allow in the text for state dependent utilities but wish to keep all along the analysis applicable to a "pure" model of portfolio selection). The second (and more important) reason is that heterogeneity of subjective probabilities appears to us to be an assumption that is most relevant, although little studied, to describe actual markets involving uncertainty, and that might generate plausible explanations of a number of so-called theoretical "puzzles" in the literature.

We show that heterogeneity in the distribution of individual subjective probabilities does generate aggregate expenditures for so-called "Arrow-Debreu" securities that are less sensitive to variations of their prices when income are price-independent. For this, heterogeneity has to be highest among agents who have the largest relative degrees of risk aversion. A key ingredient for this finding will be a relation that links the partial derivatives of individual expenditures with respect to asset prices and with respect to subjective probabilities, which will allow us to apply the methods developped for the rescaling transformation mentioned earlier even though it is not available here. The consequences of this fact will be that, in an asset exchange competitive economy where incomes are the market values of the agents' initial assets endowments, aggregate expenditures for Arrow-Debreu securities will be gross substitutes, implying in particular unicity and stability of equilibrium, and the weak axiom of revealed preferences in the aggregate.

We consider in Section 1 an individual one-period portfolio selection problem, and study how variations of subjective probabilities affect the optimal choice. We show in Section 2 how heterogeneity of subjective probabilities among people who have a large risk aversion, make aggregate expenditures less sensitive to prices. The implications for competitive exchange equilibrium in assets markets are study in Section 3.

1. Portfolio selection

We consider a standard one-period portfolio selection problem. An individual has a current income $b > 0$ that he has to invest in financial assets indexed by $j = 1, \ldots, n$. A unit of asset j generates income d_{sj} (in units of account or in kind) tomorrow in various states of the world s. If x_j is the number of units of asset j purchased, and p_j the unit price of that asset, the agent's current budget constraint is $\sum_j p_j x_j = b$. A portfolio $x = (x_j)$ generates the income $y_s = \sum_j d_{sj} x_j$ in each state. We impose the constraint that income in each state s has to be nonnegative, i.e. $y_s \geq 0$, and assume that the agent seeks to maximize the expected utility of his income, i.e. $\sum_s \pi_s u_s(y_s)$, where π_s is the subjective probability he attaches to state s. We suppose throughout

(1.*a*). *Each (possibly state dependent) von Neumann-Morgenstern utility* $\mathrm{u_s(y)}$ *is defined and continuous for* $\mathrm{y} \geq 0$, *twice continuously differentiable for* $\mathrm{y} > 0$, *with* $\mathrm{u_s'(y)} > 0$, $\mathrm{u_s''(y)} < 0$. *Moreover,* $\lim_{\mathrm{y}\to 0} \mathrm{u_s'(y)} = +\infty$ *and* $\lim_{\mathrm{y}\to +\infty} \mathrm{u_s'(y)} = 0$.

Although we shall interpret primarily the model in terms of a standard portfolio selection problem, where utility is usually supposed to be independent of the state, we allow for state dependent utilities because the analysis can also be applied to insurance problems, where the realization of some events (e.g. disease) may affect individual welfare. The assumption that marginal utility of income goes to $+\infty$ (or to 0) as income goes to 0 (or to $+\infty$) is made here for convenience in order to avoid corner solutions, and can presumably (although we did not check details) be replaced by appropriate Inada conditions.

We assume complete markets, i.e. the payoff matrix $D = (d_{sj})$ is $n \times n$ and has full rank with $n \geq 2$. As is well known, absence of arbitrage opportunities implies the existence of state prices $q_s > 0$ such that each asset j is valued according to $\sum_s q_s d_{sj} = p_j$. Then $y_s \geq 0$ can be interpreted as the demand for the *Arrow-Debreu security* corresponding to state s, which yields one unit of income in state s and none otherwise, while $q_s > 0$ is the unit price of that security. The choice of a portfolio $x = (x_j)$ is then equivalent to choosing a vector of demands $y = (y_s)$ for Arrow-Debreu securities so as to maximize expected utility under the budget constraint $\sum_s q_s y_s = b$.

The aim of this section is to investigate a few properties of the *individual demand for Arrow-Debreu securities* $y_s(q, b, \pi)$, or of the corresponding *expenditure* $w_s(q, b, \pi) = q_s y_s(q, b, \pi)$, in particular as a function of the subjective probabilities $\pi = (\pi_s)$. More specifically, we may without loss of generality relax the constraint $\sum_s \pi_s = 1$ imposed on probabilities, and suppose that the components of the vector π can move freely independently of each other. Optimal expenditures are then homogenous of degree 0 with respect to the *vector* π. Our objective is to establish a few simple relationships between the effect on these expenditures of a change in the components of the vector π, on the one hand, and of a variation in the state prices $q = (q_s)$ or income $b > 0$, on the other.

To see the basis for such relationships, we may look at the case where each (possibly state dependent) utility has a constant elasticity. The agent's problem can then be reformulated as the maximization of $\sum_s \pi_s u_s(w_s/q_s)$, subject to the constraint $\sum_s w_s = b$, where

$$\pi_s u_s(w_s/q_s) = \frac{\pi_s}{1-\rho_s}(\frac{w_s}{q_s})^{1-\rho_s} \tag{1.1}$$

when the coefficient of relative risk aversion $\rho_s \neq 1$ (the case $\rho_s = 1$, where $u_s(y) = \log y$, is as usual dealt with by continuity). It is then clear that changing q_s into λq_s, is equivalent to changing π_s into $\pi_s/\lambda^{1-\rho_s}$: the effect of a one per cent increase of the state price q_s is equal to effect of a $(1-\rho_s)$ per cent decrease of the component π_s of the vector π. The general specification $u_s(y_s)$ of the utilities in fact generates locally the same picture.

Lemma 1.1. *Under assumption* (1.a), *let the components* $\pi_s > 0$ *of the vector* π *free to move independently of each other, without imposing* $\sum_s \pi_s = 1$. *Optimal expenditures* $w_s(q, b, \pi) > 0$ *on Arrow-Debreu securities are then continuously diffrentiable, homogenous of degree* 1 *in* (q, b), *of degree* 0 *in* π, *and satisfy the budget identity* $\sum_s w_s(q, b, \pi) \equiv b$. *Moreover, if one defines the coefficients of relative risk aversion* $\rho_s(y) = -yu_s''(y)/u_s'(y)$, *and uses the notation* $\rho_s \equiv \rho_s(y_s(q, b, \pi))$, $w_s \equiv w_s(q, b, \pi)$,

$$\frac{\partial w_r}{\partial \log q_s} + (1-\rho_s)\frac{\partial w_r}{\partial \log \pi_s} \equiv 0, \tag{1.2}$$

$$\frac{\partial w_r}{\partial \log \pi_s} \equiv \frac{w_r}{\rho_r}(\delta_{rs} - \frac{\partial w_s}{\partial b}), \tag{1.3}$$

where δ_{rs} *is the Kronecker symbol, i.e.* $\delta_{rs} = 1$ *if* $r = s$, *and* $= 0$ *otherwise, and*

$$\frac{\partial w_s}{\partial b} \equiv \frac{w_s/\rho_s}{\sum_i w_i/\rho_i}. \tag{1.4}$$

Proof. Optimal expenditures are interior ($w_s > 0$), and are characterized by the budget constraint and the first order condition, for every $r \neq s$

$$\frac{\pi_s}{q_s}u_s'(\frac{w_s}{q_s}) = \frac{\pi_r}{q_r}u_r'(\frac{w_r}{q_r}). \tag{1.5}$$

The fact that optimal expenditures are continuously differentiable follows from the implicit function theorem. To prove (1.2), one remarks that (1.5) is an identity when w_r, w_s are considered as functions of (q, b, π). By differentiating (the logarithm of) (1.5) with respect to $\log \pi_s$ and $\log q_s$, one gets for every $r \neq s$

$$\frac{\rho_s}{w_s} \frac{\partial w_s}{\partial \log \pi_s} - 1 = \frac{\rho_r}{w_r} \frac{\partial w_r}{\partial \log \pi_s},$$

$$\frac{\rho_s}{w_s} \frac{\partial w_s}{\partial \log q_s} + 1 - \rho_s = \frac{\rho_r}{w_r} \frac{\partial w_r}{\partial \log q_s}.$$

This implies, by adding the first line, multiplied by $1-\rho_s$, to the second, that

$$\frac{\rho_s}{w_s}[\frac{\partial w_s}{\partial \log \pi_s}(1-\rho_s) + \frac{\partial w_s}{\partial \log q_s}] = \frac{\rho_r}{w_r}[\frac{\partial w_r}{\partial \log \pi_s}(1-\rho_s) + \frac{\partial w_r}{\partial \log q_s}]. \quad (1.6)$$

Now, because of the budget identity, one has

$$\sum_r [\frac{\partial w_r}{\partial \log \pi_s}(1-\rho_s) + \frac{\partial w_r}{\partial \log q_s}] = 0. \quad (1.7)$$

But (1.6) implies that all bracketed elements of this sum have the same sign. Hence each must be equal to 0, which is (1.2). Relations (1.3) and (1.4) are proved along similar lines. Details are left to the reader. Q.E.D.

2. Market demand

The previous statement reformulates in part a few standard "textbook" results. Owing to the separability of the utility function, Arrow-Debreu securities are normal commodities ($\partial w_r/\partial b > 0$). Individual demand for the security corresponding to state s increases if the subjective probability of other states goes down, or when the probability of that state goes up ($\partial w_r/\partial \log \pi_s < 0$ if $r \neq s$, > 0 when $r = s$). These properties are additive, and do carry over to market demand when agents have heterogenous characteristics, if all incomes or all subjective probabilities of some state move in the same direction. Yet this does not generate much structure for aggregate demand.

One property we are much interested in, is *gross substitutability*. Statement (1.2) above confirms another standard textbook result (see e.g. A. Mas-Colell, M.D. Whinston and J.R. Green (1995), p.612), namely that individual demand for Arrow-Debreu securities in complete markets does have this property (i.e. $\partial w_r/\partial \log q_s > 0$ for $r \neq s$), if the agent is an expected utility maximizer and if he has a low risk aversion ($\rho_s < 1$ for all s). Again gross substitutability is an additive property, and it does carry over to aggregate demand if *all* agents have a low risk aversion. But this condition is very strong.

We show in this Section, in particular, that a significant heterogeneity in the subjective probabilities of individual agents, tends to make aggregate expenditures on Arrow-Debreu securities less sensitive to state prices. For this, the key relation will be (1.2) above, which links partial derivatives of individual expenditures with respect to state prices, to their partial derivatives with respect to probabilities : this will allow us to use the methods developped in J.M. Grandmont (1992) for related aggregation issues. As a consequence, as will be shown in the next Section, aggregate demand for Arrow-Debreu securities in an exchange asset economy, where individual incomes are given by the market value of the agents' individual endowments, will satisfy the gross substitutability property when the heterogeneity of subjective probabilities is significant.

We define a population of agents as follows. There is first a set A of "types", which we take, in order to fix ideas, as a *separable metric space*. A type a in A is characterized by the (possibly state dependent) utilities $u_{as}(y)$, satisfying assumption (1.a) in a continuous way with respect to a.

(2.a). *For each state* s, *and each type* a *in* A, *the utility function* $\mathrm{u_{as}(y)}$ *satisfies assumption* (1.a). *Moreover,* $\mathrm{u_{as}(y)}$, *as well as its partial derivatives, are jointly continuous in* (a, y).

We consider the same market situation described in the previous Section. In particular, an agent of type a has a price independent income $b_a > 0$. The distribution of the population over the different types is described by a probability distribution μ on A. To simplify matters, we shall assume that *μ has a compact support* (this can be dispensed with, at the cost of some technicalities). We assume

(2.b). *Each type* a *in* A *has an income* $\mathrm{b_a} > 0$ *that is continuous in* a. *Aggregate (more precisely, average) income is* $\int_{\mathrm{A}} \mathrm{b_a}\, \mu(\mathrm{da}) = \overline{\mathrm{b}}$.

We assume that all agents face the same price system p_j (markets are competitive), and that *they all anticipate the same payoff matrix* $D = (d_{sj})$. In the absence of arbitrage opportunities, this implies that they all face the same implicit system of state prices $q = (q_s)$. Agents may differ, however, on the subjective probabilities $\pi = (\pi_s)$ they assign to each state. An agent of type a endowed with the probabilities $\pi = (\pi_s)$ will in effect choose expenditures on Arrow-Debreu securities, noted $w_{as}(q, b_a, \pi)$, as explained in the previous Section. Under the above assumptions these expenditures, as well as their partial derivatives, are jointly continuous in (a, q, b_a, π).

From the previous Section, the relevant variables are not so much the probabilities themselves, but rather their logarithms. We choose accordingly to parameterize subjective probabilities as follows

$$\pi_s(\gamma) = e^{\gamma_s}/(\sum_r e^{\gamma_r}). \tag{2.1}$$

Since probabilities do sum to 1 (or equivalently, because individual expenditures $w_{as}(q, b_q, \pi)$ are homogenous of degree 0 in π), we have to normalize the parameters γ_s in some way. We choose, without loss of generality, to set $\gamma_n = 0$, and to describe the distribution of subjective probabilities among people of type a by a probability distribution over the vector of remaining parameters.

(2.c). $\gamma_n = 0$. *The distribution of subjective probabilities among agents of type* a *is represented by a probability distribution over the vector of parameters* $\gamma = (\gamma_1, \ldots, \gamma_{n-1})$ *in* R^{n-1}, *with a density noted* $f(\gamma|a)$, *that is jointly continuous in* (γ, a). [1]

We have then

$$w_{as}(q, b_a, \pi(\gamma)) = w_{as}(q, b_a, \gamma_1, \ldots, \gamma_{n-1}, 0) \stackrel{\text{def}}{=} w_{as}(q, b_a, \gamma). \tag{2.2}$$

With this notation, *conditional aggregate expenditures on Arrow-Debreu securities of all agents of type* a, are given by

$$W_{as}(q, b_a) = \int w_{as}(q, b_a, \gamma) f(\gamma|a)\, d\gamma, \tag{2.3}$$

while total aggregate expenditures are obtained by summation over all types

$$W_s(q) = \int_A W_{as}(q, b_a)\, \mu(da). \tag{2.4}$$

Aggregate price effects

We wish to show that a significant heterogeneity, measured in a specific sense, in the distribution of subjective probabilities of type a agents, will tend to make their aggregate expenditures on Arrow-Debreu securities $W_{as}(q, b_a)$ less sensitive to variations of the state prices system q.

The precise measure of heterogeneity we need is provided in the following assumption.

(2.*d*). *For each type* a, *the density function* $f(\gamma|a)$ *is continuously differentiable, and its partial derivatives are jointly continuous in* (γ, a). *Moreover, for* $s = 1, \ldots, n-1$,

$$\int |(\frac{\partial f}{\partial \gamma_s}(\gamma|a)|\, d\gamma = m_{as} < +\infty,$$

[1] This formulation assumes a continuum of people for each type a. It may be the result, for instance, of assuming randomly distributed vectors γ in the population of agents of type a, appealing to the law of large numbers and going to the limit. The case of finite populations can be handled by the methods introduced by J. Quah(1997).

while for $\mathrm{s} = \mathrm{n}$,

$$\int |\sum_{\mathrm{s}=1}^{\mathrm{n}-1} \frac{\partial \mathrm{f}}{\partial \gamma_{\mathrm{s}}}(\gamma|\mathrm{a})|\, d\gamma = \mathrm{m}_{\mathrm{an}}.$$ [2]

The coefficients m_{as} measure the total variations of the density along various directions. By a large heterogeneity, we mean small coefficients m_{as}. In that case, the distribution not only has a significant variance, but its density is "flat", i.e. close (in a C^1 fashion) to a uniform distribution. Note that the last coefficient introduced for $s = n$, for reasons of symmetry, satisfies $m_{an} < \sum_{s=1}^{n-1} m_{as}$, and is accordingly small whenever the other coefficients are.

Under our assumptions, aggregate expenditures of all agents of type a are continuously differentiable, and by Leibniz's rule

$$\frac{\partial W_{ar}}{\partial \log q_s}(q, b_a) = \int \frac{\partial w_{ar}}{\partial \log q_s}(q, b_a, \gamma) f(\gamma|a)\, d\gamma.$$

We next apply (1.2) of Lemma 1.1 to each element under the integral sign. Let ρ_{am} and ρ_{aM} be the infimum and the supremum (supposed to be finite) of the degrees of relative risk aversion, $\rho_{as}(y) = -y u''_{as}(y)/u'_{as}(y)$. If $r \neq s$, we know that $\partial w_{ar}/\partial \log \pi_s < 0$. In that case, (1.2), together with (2.3), implies

$$(1 - \rho_{aM})B \leq \frac{\partial W_{ar}}{\partial \log q_s}(q, b_a) \leq (1 - \rho_{am})B \tag{2.5}$$

where $B > 0$ is given by, for $s \neq n$

$$B = -\int \frac{\partial w_{ar}}{\partial \gamma_s}(q, b_a, \gamma) f(\gamma|a)\, d\gamma, \tag{2.6}$$

whereas for $s = n$, $B > 0$ is equal to (by using homogeneity of degree 0 of $w_{ar}(q, b_a, \pi)$ with respect to the vector π, and the corresponding Euler identity)

$$B = -\int \sum_{s=1}^{n-1} \frac{\partial w_{ar}}{\partial \gamma_s}(q, b_a, \gamma) f(\gamma|a)\, d\gamma. \tag{2.7}$$

By integrating by parts (2.6), one gets

$$B = \int w_{ar}(q, b_a, \gamma) \frac{\partial f}{\partial \gamma_s}(\gamma|a)\, d\gamma,$$

while the same procedure applied to (2.7) yields

$$B = \int w_{ar}(q, b_a, \gamma) \sum_{s=1}^{n-1} \frac{\partial f}{\partial \gamma_s}(\gamma|a)\, d\gamma.$$

[2] The continuity assumptions made in (2.*c*), (2.*d*), are made to simplify the exposition. The analysis applies equally well to discontinuous (e.g. uniform) distributions, at the cost of some technicalities, see J. Quah (1997).

Therefore in all cases $s = 1, \ldots, n$, in view of assumption (2.d), $B > 0$ is bounded above by $b_a m_{as}$.

When $r = s$, we know that $\partial w_{ar}/\partial \log \pi_r > 0$. So the above argument applies to $(-\partial W_{ar}(q, b_a)/\partial \log q_r)$.

Proposition 2.1. *Assume (2.a), (2.b), (2.c), (2.d). Aggregate expenditures for Arrow-Debreu securities of all agents of type* a, *i.e.* $\mathrm{W}_{\mathrm{as}}(\mathrm{q}, \mathrm{b}_{\mathrm{a}})$, *are continuously differentiable. Let* $0 \leq \rho_{\mathrm{am}}$ *and* $\rho_{\mathrm{aM}} < +\infty$ *be the infimum and the supremum of the degrees of relative risk aversion* $\rho_{\mathrm{as}}(\mathrm{y}) = -\mathrm{y}\mathrm{u}''_{\mathrm{as}}(\mathrm{y})/\mathrm{u}'_{\mathrm{as}}(\mathrm{y})$, *over income* $\mathrm{y} > 0$ *and the state* s. *Then for every* $\mathrm{r} \neq \mathrm{s}$

$$(1 - \rho_{\mathrm{aM}})\mathrm{B} \leq \frac{\partial \mathrm{W}_{\mathrm{ar}}}{\partial \log \mathrm{q}_{\mathrm{s}}}(\mathrm{q}, \mathrm{b}_{\mathrm{a}}) \leq (1 - \rho_{\mathrm{am}})\mathrm{B}, \tag{2.8}$$

where $0 < \mathrm{B} \leq \mathrm{b}_{\mathrm{a}}\mathrm{m}_{\mathrm{as}}$. *When* $\mathrm{r} = \mathrm{s}$, *the above inequalities are satisfied by* $(-\partial \mathrm{W}_{\mathrm{ar}}(\mathrm{q}, \mathrm{b}_{\mathrm{a}})/\partial \log \mathrm{q}_{\mathrm{r}})$.

The foregoing result does imply that aggregate expenditures become less sensitive to variations of the state prices, when heterogeneity of subjective probabilities is large among agents of type a, i.e. when the coefficients m_{as} are small. In fact, the result is "additive", it is preserved when aggregating over all types present in the population. Total aggregate expenditures on Arrow-Debreu securities become less sensitive to state prices when heterogeneity on probabilities is large (the coefficients m_{as} are small) for agents who have high incomes b_a and have large risk aversions.

Corollary 2.1. *Let* η_{a} *be the maximum of* $|1 - \rho_{\mathrm{am}}|$ *and of* $|1 - \rho_{\mathrm{aM}}|$. *Then for every type* a *present in the population and every states* r, s

$$\left|\frac{\partial \mathrm{W}_{\mathrm{ar}}}{\partial \log \mathrm{q}_{\mathrm{s}}}(\mathrm{q}, \mathrm{b}_{\mathrm{a}})\right| \leq \eta_{\mathrm{a}}\mathrm{b}_{\mathrm{a}}\mathrm{m}_{\mathrm{as}}. \tag{2.9}$$

For total aggregate expenditures,

$$\left|\frac{\partial \mathrm{W}_{\mathrm{r}}}{\partial \log \mathrm{q}_{\mathrm{s}}}(\mathrm{q})\right| \leq \int_{\mathrm{A}} \eta_{\mathrm{a}}\mathrm{b}_{\mathrm{a}}\mathrm{m}_{\mathrm{as}}\, \mu(\mathrm{da}). \tag{2.10}$$

In particular, let $\eta_a m_{as}$ *be bounded above by* m_{s} *in the support of* μ. *Then*

$$\left|\frac{\partial \mathrm{W}_{\mathrm{r}}}{\partial \log \mathrm{q}_{\mathrm{s}}}(\mathrm{q})\right| \leq \bar{\mathrm{b}}\mathrm{m}_{\mathrm{s}}. \tag{2.11}$$

3. Exchange equilibrium

We now study the implications of our findings for existence, uniqueness and stability of competitive equilibrium in an assets exchange economy, where the income b_a of every agent is the market value of his initial endowment of assets. This will be achieved by a straight adaptation of the methods and results described in Grandmont (1992).

We retain all the assumptions of the previous Section. We suppose further that every individual of type a is endowed with a vector $\overline{x}_a$ of initial assets, so that his income is $b_a = \sum_j p_j \overline{x}_{aj}$. Under our assumptions (people expect the same payoff matrix D, complete markets, absence of arbitrage opportunities), this is equivalent to suppose that the agent has an initial endowment of Arrow-Debreu securities $\omega_a = D\overline{x}_a$, whose market value is $b_a = q.\omega_a$. We assume that the endowment ω_a is nonnegative, nonzero, continuous in a, and that all Arrow-Debreu securities are present in positive amounts in the aggregate.

(3.*a*). *The assumptions* (2.*a*), (2.*b*), (2.*c*), (2.*d*) *are retained. Every type* a *in* A *is initially endowed with a vector of Arrow-Debreu securities* $\omega_{\mathrm{a}} \geq 0$, $\omega_{\mathrm{a}} \neq 0$, *that depends continuously on* a. *The aggregate endowment has all its components positive, i.e.* $\int_{\mathrm{A}} \omega_{\mathrm{a}}\, \mu(\mathrm{da}) = \overline{\omega} >> 0$.

Conditional aggregate expenditures on Arrow-Debreu securities of all agents of type a are then given by replacing b_a by $q.\omega_a$ in (2.3)

$$W_{as}(q, q.\omega_a) = \int w_{as}(q, q.\omega_a, \gamma) f(\gamma|a)\, d\gamma, \tag{3.1}$$

and total aggregate expenditures are obtained by performing the same operation in (2.4)

$$W_s(q) = \int_A W_{as}(q, q.\omega_a)\, \mu(da). \tag{3.2}$$

A *competitive exchange equilibrium* is then defined as a vector of positive prices q^* for Arrow-Debreu securities, such that $W_s(q^*) = q_s^* \overline{\omega}_s$ for every state $s = 1, \ldots, n$. The corresponding equilibrium price system $p^* = (p_j^*)$ for the original assets is then given as usual by using the payoff matrix, $p_j^* = \sum_s q_s^* d_{sj}$.

We wish to show here that when there is a significant heterogeneity in the agents' subjective probabilities, i.e. when the parameters (m_{as}) introduced in assumption (2.*d*) are small, equilibrium is unique, stable in any tâtonnement process, and that the weak axiom of revealed preference is satisfied in the aggregate, as between an equilibrium price vector q^*.

To achieve this program, we would like to argue that with enough heterogeneity in subjective probabilities, Arrow-Debreu securities are gross substitutes in the aggregate, i.e. $\partial W_r(q)/\partial \log q_s > 0$ for every $r \neq s$. Now, it is clear from (3.2) that $\partial W_r(q)/\partial \log q_s$ is the sum of two terms

$$\begin{aligned} \frac{\partial W_r(q)}{\partial \log q_s} &= \int_A \frac{\partial W_{ar}}{\partial \log q_s}(q, q.\omega_a)\, \mu(da) \\ &+ \int_A \frac{q_s.\omega_{as}}{q.\omega_a} \frac{\partial W_{ar}}{\partial \log b}(q, q.\omega_a)\, \mu(da) \end{aligned} \tag{3.3}$$

(we remind the reader that we assume μ to have compact support, so that differentiation under the integral sign is indeed legitimate). From Proposition 2.1 and Corollary 2.2, the first term in (3.3) can be made small when heterogeneity of subjective probabilities is significant. To get gross substitutability, we need that the second term, which we know to be positive (with separable utilities, Arrow-Debreu securities are normal commodities, see (1.4), and this property aggregates from the micro to the macrolevel), to be bounded away from 0, independently of the degree of heterogeneity of the agents' subjective probabilities. There are several ways to deal with this issue; we shall proceed here, as in Grandmont (1992), by making assumptions ensuring , loosely speaking, that aggregate budget shares for Arrow-Debreu securities are bounded away from 0. The same type of assumptions will also guarantee that equilibrium security prices q^* do remain bounded away from 0, again independently of degree of heterogeneity of subjective probabilities.

(3.*b*). *There is a set* A_0 *of types who have logarithmic expected utility* $\sum_{\mathrm{s}} \pi_{\mathrm{as}} \log \mathrm{y}_{\mathrm{s}}$. *Let mean subjective probabilities among agents of type* a *be defined as* $\overline{\pi}_{\mathrm{as}} = \int \pi_{\mathrm{s}}(\gamma)\mathrm{f}(\gamma|\mathrm{a})\,\mathrm{d}\gamma$. *Then there exists* $0 < \varepsilon_{\mathrm{s}} < 1$ *for every state* s, *such that*

$$\int_{\mathrm{A}_0} \overline{\pi}_{\mathrm{ar}}(\mathrm{q}_{\mathrm{s}}\omega_{\mathrm{as}})\,\mu(\mathrm{da}) \geq \varepsilon_{\mathrm{r}}(\mathrm{q}_{\mathrm{s}}\overline{\omega}_{\mathrm{s}}) \tag{3.4}$$

The above (admittedly crude) assumption is especially tailored to enable us to apply directly the results of Grandmont (1992). Indeed,

Theorem 3.1. *Let assumptions* (3.*a*), (3.*b*) *hold. For every type* a, *let* $\eta_{\mathrm{a}} < +\infty$ *be the supremum of* $|1 - \rho_{\mathrm{s}}(\mathrm{y})|$ *for all* $\mathrm{y} > 0$ *and all states* s, *and assume that* $\eta_{\mathrm{a}}\mathrm{m}_{\mathrm{as}}$ *is bounded above by* m_{s} *in the support of* μ. *Then*

(1) *There exists an equilibrium price system* $\mathrm{q}^* >> 0$ *for Arrow-Debreu securities. It satisfies* $\mathrm{q}^*\overline{\omega}_{\mathrm{s}} \geq \varepsilon_{\mathrm{s}}(\mathrm{q}^*.\overline{\omega})$.

Let $\mathrm{GS}(\mathrm{m},\varepsilon)$ *be the convex cone of prices* $\mathrm{q} >> 0$ *satisfying*

$$\mathrm{q}_{\mathrm{s}}\overline{\omega}_{\mathrm{s}}(\varepsilon_{\mathrm{r}} - \sum_{\mathrm{i}} \mathrm{m}_{\mathrm{i}}) > \mathrm{m}_{\mathrm{s}}(\mathrm{q}.\overline{\omega}), \quad \textit{all} \ \mathrm{r} \neq \mathrm{s}. \tag{3.5}$$

(2) *Arrow-Debreu securities are gross substitutes, i.e.* $\partial \mathrm{W}_{\mathrm{r}}(\mathrm{q})/\partial \log \mathrm{q}_{\mathrm{s}} > 0$ *for* $\mathrm{r} \neq \mathrm{s}$, *when the price system* q *lies in* $\mathrm{GS}(\mathrm{m},\varepsilon)$.

(3) *Assume*

$$\varepsilon_{\mathrm{s}}(\varepsilon_{\mathrm{r}} - \sum_{\mathrm{i}} \mathrm{m}_{\mathrm{i}}) > \mathrm{m}_{\mathrm{s}}, \quad \textit{all} \ \mathrm{r} \neq \mathrm{s}. \tag{3.6}$$

Then $\mathrm{GS}(\mathrm{m},\varepsilon)$ *contains the set of prices* $\mathrm{q} >> 0$ *such that* $\mathrm{q}_{\mathrm{s}}\overline{\omega}_{\mathrm{s}} \geq \varepsilon_{\mathrm{s}}(\mathrm{q}.\overline{\omega})$, *and the equilibrium price vector* q^* *is unique (up to multiplication by a scalar).*

(4) *Let aggregate excess demands for Arrow-Debreu securities be defined as* $\mathrm{Z_s(q) = (W_s(q) - q_s\overline{\omega}_s)/q_s}$ *for every state* s. *Then*

$$\varepsilon_{\mathrm{s}}^2(\varepsilon_{\mathrm{r}} - \sum_{\mathrm{i}} \mathrm{m_i}) > \mathrm{m_s}, \ \textit{all}\ \mathrm{r} \neq \mathrm{s}, \tag{3.7}$$

implies that the weak axiom of revealed preference holds in the aggregate, i.e. $\mathrm{q^*.Z(q) > 0}$, *as between an equilibrium price system* $\mathrm{q^*}$ *and any other price vector* $\mathrm{q >> 0}$ *that is not colinear to* $\mathrm{q^*}$.

Proof. (1) Existence of an equilibrium price vector q^* under our assumptions is a standard result. On the other hand,

$$W_s(q) \geq \int_{A_0} W_{as}(q, q.\omega_a)\ \mu(da) = \int_{A_0} \overline{\pi}_{as}(q.\omega_a)\ \mu(da).$$

Hence, in view of (3.4), $W_s(q) \geq \varepsilon_s(q.\overline{\omega})$, and one can apply Proposition 3.2 of Grandmont (1992).

To prove (2), (3), (4), we first look for a lower bound for $\partial W_r(q)/\partial \log q_s$, by using (3.3). In view of Corollary 2.2, the first integral in (3.3) is bounded below by $-m_s(q.\overline{\omega})$. To bound the second term, we remark that $W_{as}(q, b_a)$ is homogenous of degree 1 in prices and income. Using Euler's identity and Corollary 2.2, we conclude that

$$|\frac{\partial W_{ar}}{\partial \log b}(q, b_a) - W_{ar}(q, b_a)| \leq b_a \eta_a \sum_i m_{ai} \leq b_a \sum_i m_i$$

in the support of μ. Therefore the second integral in (3.3) is bounded below

$$\int_A q_s \omega_{as}[W_{ar}(q, q.\omega_a)/q.\omega_a]\ \mu(da) - (q_s\overline{\omega}_s)\sum_i m_i.$$

The above integral is itself bounded below by

$$\int_{A_0} (q_s\omega_{as})\overline{\pi}_{ar}\ \mu(da) \geq \varepsilon_r(q_s\overline{\omega}_s),$$

by using (3.4) in assumption (3.*b*). Thus a lower bound for $\partial W_r(q)/\partial \log q_s$ is $q_s\overline{\omega}_s(\varepsilon_r - \sum_i m_i) - m_s(q.\omega_s)$. It is then clear that aggregate expenditures on Arrow-Debreu securities have the gross substitutability property for security prices q in $GS(m, \varepsilon)$. The other statements (3), (4) follow immediately, exactly as in Theorem 3.3 in Grandmont (1992). Q.E.D.

The above result shows how a significant heterogeneity in the agents' subjective probabilities generates aggregate demands for Arrow-Debreu securities that satisfy the gross substitutability property and the weak axiom of revealed preferences. This heterogeneity leads also to increasing stability of the (unique) equilibrium under any kind of tâtonnement process : see Corollary 3.4 in Grandmont (1992), which applies without any change here.

4. Conclusion

We have shown in this paper how heterogeneity in subjective probabilities can lead in complete competitive asset markets, to aggregate demands for Arrow-Debreu securities that are gross substitutes, with the consequences that exchange equilibrium is unique and stable in any tatônnement process, and that the weak axiom of revealed preferences is satisfied in the aggregate. To get this result, heterogeneity had to be highest among people who displayed large relative degrees of risk aversion. It remains to be seen whether the methods and results of this paper can be made to bear on other models involving uncertainty such as incomplete asset markets, or markets for insurance.

References

1. Grandmont, J.-M.: Transformations of the commodity Space, behavioral heterogeneity and the aggregation problem. Journal of Economic Theory **57**, 1-35 (1992)
2. Mas-Colell, A., Whinston, M.D., Green, J.R.: Microeconomic Theory. Oxford University Press 1995
3. Quah, J.: The law of demand when income is price dependent. Econometrica **65**, 1421-1442 (1997)

Adv. Math. Econ. 1, 17–37 (1999)

Advances in
MATHEMATICAL
ECONOMICS

Convergences in $L^1_X(\mu)$

Charles Castaing[1] **and Mohamed Guessous**[2]

[1] Département de Mathématiques, Case 051, Université Montpellier II, F-34095 Montpellier cedex 5, France

[2] Université Hassan II Mohamedia, Faculté des Sciences Ben M'sik, Département de Mathématiques, BP 7955 Ben M'sik, Casablanca, Maroc

Received: March 6, 1998

JEL classification: C69

Mathematics Subject Classification (1991): 46E30, 28A20, 60B12

Summary. We present new modes of convergences for bounded sequences in the space $L^1_X(\mu)$ of Bochner integrable functions over a complete probability space $(\Omega, \mathcal{F}, \mu)$ with values in Banach space X via the convergence of its truncated subsequences as well as we give several characterizations of weak compactness and conditionally weak compactness in $L^1_X(\mu)$. New results involving subsets in $L^1_X(\mu)$ which are closed in measure are obtained and also the characterizations of the Banach space X in terms of these modes of convergence.

Key words: Compactness, Komlós, Talagrand, tight.

1. Introduction

Let X be a Banach space and $L^1_X(\mu)$ the Banach space of (equivalence classes of) Lebesgue-Bochner integrable functions over a complete probability space $(\Omega, \mathcal{F}, \mu)$. Ülger-Diestel-Ruess-Shachermayer [23, 35] proved that a bounded uniformly integrable subset $\mathcal{H}$ of $L^1_X(\mu)$ is relatively weakly compact iff,

(*) given any sequence (u_n) in $\mathcal{H}$, there exists a sequence (v_n), with $v_n \in co\{u_m : m \geq n\}$ such that $(v_n(\omega))$ is weakly convergent in X for almost all $\omega \in \Omega$.

The proof given in [35] relies on a deep result due to Talagrand [34] concerning the decomposition of bounded sequences in $L^1_X(\mu)$. Díaz [24] and Benabdellah-Castaing [9, 10] describe the above "weak compactness condition" in terms of regular method of summability providing new characterizations of weak compactness and weak conditionally compactness in X and $L^1_X(\mu)$. In [9, 10], the authors gave several characterizations of the condition (*), called as *weak Talagrand property (WTP)* because of the above mentioned decomposition, and stated several properties of WTP sets and $\mathcal{R}_w(X)$-*tight* sets in $L^1_X(\mu)$. Namely by $\mathcal{R}_w(X)$ we denote the set of nonempty convex closed subsets of X such that their intersection with any closed ball is weakly compact. A subset $\mathcal{H}$ of $L^1_X(\mu)$ is $\mathcal{R}_w(X)$-tight if, for every $\varepsilon > 0$ there exists a $\mathcal{R}_w(X)$-valued measurable multifunction $\Gamma_\varepsilon : \Omega \to X$ such that $\mu(\{\omega \in \Omega : u(\omega) \notin \Gamma_\varepsilon(\omega)\}) \leq \varepsilon$ for all $u \in \mathcal{H}$. It was shown in [9, 10] that the $\mathcal{R}_w(X)$-tight property implies the WTP. Moreover, if $\mathcal{H}$ is a bounded uniformly integrable $\mathcal{R}_w(X)$-tight subset of $L^1_X(\mu)$ then $\mathcal{H}$ is relatively weakly compact [2, 15] and any sequence (u_n) in $\mathcal{H}$ admits a subsequence which weakly converges in $L^1_X(\mu)$. Using this tightness condition, several criteria of weak compactness and weak conditionally compactness were stated in [9, 10]. Recently Saadoune [31, 32] obtained a generalization of Ülger-Diestel-Ruess-Shachermayer theorem and other variants by replacing in $(*)$ the condition : $(v_n(\omega))$ *is weakly convergent* by the following : (v_n) *is weak-tight*, namely, for every $\varepsilon > 0$, there is a convex weakly compact valued measurable multifunction $\Gamma_\varepsilon : \Omega \to X$ such that:

$$\sup_n \mu(\{\omega \in \Omega : v_n(\omega) \notin \Gamma_\varepsilon(\omega)\}) \leq \varepsilon.$$

He used Komlós arguments and weak compactness property for bounded uniformly integrable weak-tight subsets in $L^1_X(\mu)$. In the same vein, Díaz [24] obtained several results related to the weak Césaro convergence (WCP) in $L^1_X(\mu)$ and also the characterization of weak compactness in $L^1_X(\mu)$ by the weak Césaro convergence. We refer to [1, 7, 16, 19, 31, 32] for other related results in the Komlós convergence.

In this paper we aim to state new results concerning the convex WTP subsets in $L^1_X(\mu)$ which are closed in measure extending some results in [13, 14, 30] and also the relationships between the WCP, WTP and WKP and other modes of convergence we introduce below. In particular we characterize the WTP for bounded subsets in $L^1_X(\mu)$ in terms of the $\sigma(L^1, L^\infty)$ convergence of the truncated sequences $(1_{\{||u_n||<n\}}u_n)_n$ $(n \in \mathbb{N}^*)$ from the sequences (u_n) in the sets into consideration and reflexive Banach spaces in terms of these modes of convergence.

We also investigate the properties of *conditionally WTP* sets in $L^1_X(\mu)$ involving several unusual modes of convergence. Finally we characterize a class of Banach spaces not containing l^1 in terms of these modes of convergence.

2. Notations and terminology

By $\mathcal{R}_w(X)$ we denote the collection of nonempty closed convex subsets of X such that their intersection with any closed ball is weakly compact. A subset

$\mathcal{H}$ of $L^1_X(\mu)$ is $\mathcal{R}_w(X)$-*tight*, if for every $\varepsilon > 0$ there exists a $\mathcal{R}_w(X)$-valued measurable multifunction $\Gamma_\varepsilon : \Omega \to X$ such that

$$\sup_{u \in \mathcal{H}} \mu(\{\omega \in \Omega : u(\omega) \notin \Gamma_\varepsilon(\omega)\}) \le \varepsilon.$$

If (f_n) is a sequence in $L^1_X(\mu)$ and if $N \in \mathbb{N}^*$, we denote by $(f_n^N)_n$ the truncated sequence associated to (f_n) and N, where $f_n^N := 1_{\{||f_n||<N\}} f_n$.

We need the following definitions. Let $\mathcal{H}$ be a subset of $L^1_X(\mu)$.

(1) $\mathcal{H}$ is said to have the *weak Talagrand property* (WTP) if, given any bounded sequence (f_n) in $\mathcal{H}$, there exists a sequence (g_n) with $g_n \in co\{f_k : k \ge n\}$ such that (g_n) weakly converges a.e in X.
(2) $\mathcal{H}$ is said to have the *weak Césaro property* (WCP) if, given any bounded sequence (f_n) in $\mathcal{H}$, there exists a subsequence (g_n) such that the Césaro sums $\frac{1}{n}\Sigma_{i=1}^n g_i$ weakly converge a.e in X.
(3) $\mathcal{H}$ is said to have the *weak Komlós property* (WKP) if, given any bounded sequence (f_n) in $\mathcal{H}$, there exist $f \in L^1_X(\mu)$ and a subsequence (g_n) such that $\frac{1}{n}\Sigma_{i=1}^n h_i$ weakly converges a.e to f in X, for each subsequence (h_n) of (g_n).

It is obvious that WKP$\Longrightarrow$WCP.

3. Weak compactness and conditionally weak compactness in $L^1_X(\mu)$

We will summarize the following weak compactness results.

Theorem 3.1. *Suppose that X is a separable Banach space, $\mathcal{H}$ is a bounded uniformly integrable $\mathcal{R}_w(X)$-tight subset of $L^1_X(\mu)$. Then $\mathcal{H}$ is relatively weakly compact and, if (u_n) is a sequence in $\mathcal{H}$, there is a subsequence of (u_n) which $\sigma(L^1, L^\infty)$ converges*[1].

Proof. See [15, Théorème 4.1]; [2, Théorème 6].

The following result is due to Ülger [35]. See also [9, 10, 23, 24] for related results.

Theorem 3.2. *Suppose that X is a Banach space and K is a subset of X. Then the two following conditions are equivalent :*

(1) *K is relatively weakly compact.*

(2) *Given any sequence (x_n) in K, there is a sequence (y_n) with $y_n \in co\{x_m : m \ge n\}$ such that (y_n) is weakly convergent.*

[1] For simplicity we denote by $\sigma(L^1, L^\infty)$ the weak topology $\sigma(L^1_X(\mu), L^\infty_{X^*}(\mu))$

The above two theorems together with Eberlein-Smulian theorem allows us to prove the following theorem [23, 35].

Theorem 3.3. *Suppose that X is a Banach space and $\mathcal{H}$ is a subset of $L^1_X(\mu)$. Then $\mathcal{H}$ is relatively weakly compact iff $\mathcal{H}$ is bounded uniformly integrable and satisfies: for any sequence (u_n) in $\mathcal{H}$ there is a sequence (v_n) with $v_n \in co\{u_m : m \geq n\}$ such that (v_n) weakly converges a.e in X.*

For more on weak compactness in $L^1_X(\mu)$, see, for example, [1, 9, 10, 18, 24, 31, 32].

Let us mention the following characterization of conditionally weakly compact subsets of $L^1_X(\mu)$ [9, 10].

Theorem 3.4. *A bounded subset $\mathcal{H}$ of $L^1_X(\mu)$ is conditionally weakly compact iff $\mathcal{H}$ is uniformly integrable and for any sequence (u_n) in $\mathcal{H}$ there is a sequence (v_n) with $v_n \in co\{u_m : m \geq n\}$ such that (v_n) is weakly Cauchy a.e in X.*

4. Uniform integrability and convergence of the truncated sequences

We will need the following two results which we include for the sake of completeness (cf. also [25, 28, 33]).

Lemma 4.1. *Let (f_n) be a bounded sequence in $L^1_X(\mu)$. Then there exists a subsequence (g_n) of (f_n) such that for each subsequence (h_n) of (g_n)*

(a) *the sequence $(1_{\{||h_n||<n\}}h_n)$ is uniformly integrable;*

(b) *the sequence $(h_n - 1_{\{||h_n||<n\}}h_n)$ converges a.e to 0 in X.*

Proof. Put $\mathrm{M} = \sup_n ||f_n||_1$. For each integer $k \geq 1$ the real valued sequence

$$(\mu(\{k-1 \leq ||f_n|| < k\}))_n$$

is bounded. Then there exist subsequences $f^1_n, f^2_n, .., f^k_n, ..$ of (f_n), where (f^{k+1}_n) is a subsequence of (f^k_n), and a sequence $(p_k)_k$ in $[0,1]$ such that

$$(j) \qquad \forall k \geq 1,\ \lim_n \mu(\{k-1 \leq ||f^k_n|| < k\}) = p_k$$

and

$$(jj) \qquad \forall k \geq 1, \forall n \geq 1,\ \mu(\{k-1 \leq ||f^k_n|| < k\}) < p_k + \frac{1}{k^3}.$$

Indeed we can extract from $(f_n)_n$ a subsequence $(f^1_{0,n})_n$ satisfying (j) for $k=1$. Since $\mu(\{0 \leq f^1_{0,n} < 1\}) \to p_1$, there exists n_0 such that

$$\forall n \geq n_0,\ \mu(\{0 \leq f^1_{0,n} < 1\}) < p_1 + 1.$$

So it suffices to set $f^1_n = f^1_{0,n+n_0-1}$ in order to have (jj) for all n. Suppose that (f^{k-1}_n) has been obtained, then using the similar arguments we can construct from (f^{k-1}_n) a subsequence (f^k_n) with the required properties.

Put $g_n = f^{n^2}_n$ and let (h_n) be a subsequence of (g_n). Then

(i) $\forall k \geq 1,\ \lim_n \mu(\{k-1 \leq ||h_n|| < k\}) = p_k$,

and

(ii) $\forall n \in \mathbb{N}^*, \forall k \in [1, n^2],\ \mu(\{k-1 \leq ||h_n|| < k\}) < p_k + \frac{1}{k^3}$.

We only need to check (ii). Since (h_n) is extracted from $(g_n)_n = (f^{n^2}_n)_n$, each h_n is of the form $h_n = f^{m^2}_m$ with $m \geq n$. Let $k \in \mathbb{N}^*$ be fixed. Let us notice that $(f^{m^2}_n)_n$ is a subsequence of $(f^k_n)_n$ when $m^2 \geq k$. So let us suppose that $n \geq \sqrt{k}$. Then $m^2 \geq k$, and so $f^{m^2}_m$ is a term f^k_l of $(f^k_n)_n$ with $l \geq m$, hence by (jj) and what has been said we deduce (ii). In this part of proof the symbol $(f^k_n)_n$ were used for notational convenience and there was no risk of confusion with the truncated sequences $(f^N_n)_n$ $(N \in \mathbb{N}^*)$ above mentioned . We have

$$\sum_{k=1}^{N} p_k = \lim_n \sum_{k=1}^{N} \mu(\{k-1 \leq ||h_n|| < k\}) \leq \mu(\Omega),$$

and

$$\sum_{k=1}^{N} (k-1)p_k = \lim_n \sum_{k=1}^{N} (k-1)\mu(\{k-1 \leq ||h_n|| < k\}) \leq M.$$

Then $\sum_{k=1}^{+\infty} p_k \leq 1$ and $\sum_{k=1}^{+\infty} (k-1)p_k \leq M$.

(a) Let $\varepsilon > 0$. As $\sum_{k=1}^{+\infty} kp_k$ converges, so is $\sum_{k=1}^{+\infty} k(p_k + \frac{1}{k^3})$, hence there exists $N \in \mathbb{N}^*$ such that

$$\sum_{k \geq N} k(p_k + \frac{1}{k^3}) < \varepsilon.$$

If $n \in \{1, .., N\}$, then $||h^n_n|| < n \leq N$, so $\{||h^n_n|| \geq N\} = \emptyset$. If $n > N$, then

$$\{||h^n_n|| \geq N\} = \bigcup_{N+1 \leq k \leq n} \{k-1 \leq ||h^n_n|| < k\},$$

and

$$\int_{\{||h^n_n|| \geq N\}} ||h^n_n||\, d\mu \leq \sum_{N+1 \leq k \leq n} k(p_k + \frac{1}{k^3}) < \varepsilon$$

so (h^n_n) is uniformly integrable.

(b) We have

$$
\begin{aligned}
\mu(\{||h_n|| \geq n\}) &= \sum_{k=n+1}^{n^2} \mu(\{k-1 \leq ||h_n|| < k\}) + \mu(\{||h_n|| \geq n^2\}) \\
&< \sum_{k=n+1}^{n^2} (p_k + \frac{1}{k^3}) + \frac{1}{n^2}||h_n||_1 \\
&\leq \sum_{k \geq n+1} p_k + \frac{1}{2n^2} + \frac{M}{n^2}.
\end{aligned}
$$

Whence by summing we get

$$
\begin{aligned}
\sum_{n\geq 1} \mu(\{||h_n|| \geq n\}) &< \sum_{n\geq 1} [\sum_{k\geq n+1} p_k + \frac{2M+1}{2n^2}] \\
&= \sum_{k\geq 2}(k-1)p_k + \sum_{n\geq 1} \frac{2M+1}{2n^2} < +\infty.
\end{aligned}
$$

Then (b) follows from Borel-Cantelli Lemma. □

Lemma 4.2. *Let (u_n) be a bounded sequence in $L^1_X(\mu)$ and (v_n) a sequence in $L^1_X(\mu)$. Suppose that, for each $k \in \mathbb{N}^*$, the sequence $(1_{\{||u_n||<k\}}u_n)_n$ weakly converges to v_k in $L^1_X(\mu)$. Then the following holds :*

(1) *(v_k) strongly and μ-a.e converges to a function v in $L^1_X(\mu)$.*

(2) *There exists a subsequence (h_n) of (u_n) such that $(1_{\{||h_n||<n\}}h_n)_n$ converges $\sigma(L^1, L^\infty)$ to v in $L^1_X(\mu)$.*

Proof. (1) Put $v_0 = 0$. It is enough to prove that the series $\sum_{k\geq 1} ||v_k - v_{k-1}||_1$ converge. For each $k \in \mathbb{N}^*$, the sequence $(u^k_n - u^{k-1}_n)_n$ converges weakly in $L^1_X(\mu)$ to $v_k - v_{k-1}$. By [27, Lemma 1] there exists $m_k \in \mathbb{N}^*$ such that

$$
||v_k - v_{k-1}||_1 \leq 2 \inf_{n\geq m_k} ||u^k_n - u^{k-1}_n||_1.
$$

Let $N \in \mathbb{N}^*$ and $n \geq \max(m_1, m_2, \dots, m_N)$. Then we have

$$
\begin{aligned}
\sum_{k=1}^{N} ||v_k - v_{k-1}||_1 &\leq 2\sum_{k=1}^{N} ||u^k_n - u^{k-1}_n||_1 \\
&\leq 2||u_n||_1 \leq 2 \sup_{p\geq 1} ||u_p||_1
\end{aligned}
$$

and therefore $\sum_{k=1}^{+\infty} ||v_k - v_{k-1}||_1 \leq 2\sup_{p\geq 1} ||u_p||_1 < +\infty$.

(2) By Lemma 4.1, there exists a subsequence (h_n) of (u_n) such that the sequence $(1_{\{||h_n||<n\}}h_n)_n$ is uniformly integrable. Let us consider $\zeta \in (L^1_X(\mu))'$ with norm ≤ 1 and $\varepsilon > 0$. By the UI of $(1_{\{||h_n||<n\}}h_n)_n$ and by (1), there is $n_0 \in \mathbb{N}$ such that

$$.\sup_n \int_{\{||h_n^n|| \geq n_0\}} ||h_n^n|| \, d\mu \leq \frac{\varepsilon}{3} \text{ and } ||v_{n_0} - v||_1 \leq \frac{\varepsilon}{3}.$$

Whence, for $n \geq n_0$, we have

$$||h_n^n - h_n^{n_0}||_1 \leq \frac{\varepsilon}{3}.$$

As $(h_n^{n_0})$ converges $\sigma(L^1, L^\infty)$ to v_{n_0}, there exist $n_1 \geq n_0$ such that

$$n \geq n_1 \Longrightarrow \langle \zeta, h_n^{n_0} - v_{n_0} \rangle \leq \frac{\varepsilon}{3}.$$

Then, for $n \geq n_1$, we have

$$\begin{aligned} \langle \zeta, h_n^n - v \rangle &\leq \langle \zeta, h_n^n - h_n^{n_0} \rangle + \langle \zeta, h_n^{n_0} - v_{n_0} \rangle + \langle \zeta, v_{n_0} - v \rangle \\ &\leq ||h_n^n - h_n^{n_0}||_1 + \langle \zeta, h_n^{n_0} - v_{n_0} \rangle + ||v_{n_0} - v||_1 \\ &\leq \varepsilon. \end{aligned}$$

□

5. Convex WTP sets closed in measure in $L^1_X(\mu)$

The characterization of weak compactness in $L^1_X(\mu)$ via uniformly integrable condition and weak Talagrand property was stated by [23, 35] in which a deep result on the decomposition of bounded sequences in $L^1_X(\mu)$ due to Talagrand [34] is involved. It is why the notion WTP was investigated in [9, 10], in particular the authors stated that a convex lower semicontinuous function for the topology of convergence in measure reaches its minimum on a bounded convex WTP subset of $L^1_X(\mu)$ which is closed in measure, thus generalizing a result stated by [13, 14, 30] in the case where X is a reflexive Banach space. As [14] contains no proofs, some of the proofs were given in [36]. In this section we continue this study and we will show some nice properties of convex WTP and closed in measure subsets in $L^1_X(\mu)$.

Proposition 5.1. *If $(\mathcal{H}_n)_{n\geq 1}$ is a sequence of bounded convex WTP subsets of $L^1_X(\mu)$ which are closed in measure and $\forall n$, $\cap_{i=1}^n \mathcal{H}_i \neq \emptyset$, then $\cap_{i=1}^\infty \mathcal{H}_i \neq \emptyset$.*

Proof. For each n, pick $f_n \in \cap_{i=1}^n \mathcal{H}_i$, then $(f_n) \subset \mathcal{H}_1$. Since $\mathcal{H}_1$ is bounded WTP, there is a sequence (f'_n) with $f'_n \in co\{f_m : m \geq n\}$ such that (f'_n) converges a.e. to a function $f \in L^1_X(\mu)$ (see, [9, 10]). But then $(f'_n)_{n\geq k} \subset \mathcal{H}_k$ for every k and each $\mathcal{H}_k$ is closed in measure, so we have that $f \in \mathcal{H}_k$ for every k, thus proving that $f \in \cap_{i=1}^\infty \mathcal{H}_i$. □

Proposition 5.2. *If $\mathcal{H}_1$ and $\mathcal{H}_2$ are convex sets in $L^1_X(\mu)$ which are closed in measure and if $\mathcal{H}_1$ is bounded and has the WTP, then the set*

$$\mathcal{H} := \{f_1 + f_2 : f_1 \in \mathcal{H}_1, f_2 \in \mathcal{H}_2\}$$

is closed in measure.

Proof. We need to prove that if (u_n) is a sequence in $\mathcal{H}$ which converges a.e to a function $u \in L^1_X(\mu)$, then $u \in \mathcal{H}$. By definition of $\mathcal{H}$, we have that $u_n = f_n + g_n$ with $f_n \in \mathcal{H}_1$ and $g_n \in \mathcal{H}_2$. Since $\mathcal{H}_1$ is bounded and WTP set, there is a sequence (f'_n) with $f'_n \in co\{f_k : k \geq n\}$ such that (f'_n) converges a.e to a function $f \in L^1_X(\mu)$ (see, [9, 10]). Each f'_n has the form $f'_n = \Sigma_{i=n}^{\nu_n} \lambda_i^n f_i$ with $\lambda_i^n \geq 0$, $\Sigma_{i=n}^{\nu_n} \lambda_i^n = 1$. Now it is obvious that the sequence $(g'_n) = (\Sigma_{i=n}^{\nu_n} \lambda_i^n g_i)_n$ converges a.e to $g = u - f \in L^1_X(\mu)$. Since $\mathcal{H}_1$ and $\mathcal{H}_2$ are convex and closed in measure, $f \in \mathcal{H}_1$ and $g \in \mathcal{H}_2$. Thus $u = f + g \in \mathcal{H}$. □

The following is a minimisation result.

Proposition 5.3. *If $\mathcal{H}_1$ and $\mathcal{H}_2$ are convex WTP sets in $L^1_X(\mu)$ which are closed in measure and if $\mathcal{H}_1$ is bounded, then there exist $f \in \mathcal{H}_1$ and $g \in \mathcal{H}_2$ such that*

$$||f - g||_1 = \inf\{||u - v||_1 : u \in \mathcal{H}_1,\, v \in \mathcal{H}_2\}.$$

Proof. Let us set $m := \inf\{||u-v||_1 : u \in \mathcal{H}_1,\, v \in \mathcal{H}_2\}$. There exist a sequence (f_n) in $\mathcal{H}_1$ and a sequence (g_n) in $\mathcal{H}_2$ such that $m = \lim_{n\to\infty} ||f_n - g_n||_1$. Since $\mathcal{H}_1$ is bounded WTP, there is a sequence (f'_n) with $f'_n \in co\{f_k : k \geq n\}$ such that (f'_n) converges a.e. to a function $f \in L^1_X(\mu)$ (see, [9, 10]). Each f'_n has the form $f'_n = \Sigma_{i=n}^{\nu_n} \lambda_i^n f_i$ with $\lambda_i^n \geq 0$, $\Sigma_{i=n}^{\nu_n} \lambda_i^n = 1$. Since (g_n) is norm-bounded in $L^1_X(\mu)$ (because $||g_n||_1 \leq ||f_n - g_n||_1 + ||f_n||_1$) and $\mathcal{H}_2$ is convex WTP, $(g'_n)_n = (\Sigma_{i=n}^{\nu_n} \lambda_i^n g_i)_n$ is a bounded WTP sequence. Hence there is a sequence $(\widetilde{g}_n)$ with $\widetilde{g}_n \in co\{g'_k : k \geq n\}$ such that $(\widetilde{g}_n)$ converges a.e to a function $g \in L^1_X(\mu)$ (see, [9, 10]). Each $\widetilde{g}_n$ has the form $\widetilde{g}_n = \Sigma_{i=n}^{\beta_n} \mu_i^n g'_i$ with $\mu_i^n \geq 0$, $\Sigma_{i=n}^{\beta_n} \mu_i^n = 1$. Let $(\widetilde{f}_n)_n = (\Sigma_{i=n}^{\beta_n} \mu_i^n f'_i)_n$. Then it is clear that

$$m = \lim_{n\to\infty} ||f_n - g_n||_1 = \lim_{n\to\infty} ||f'_n - g'_n||_1 = \lim_{n\to\infty} ||\widetilde{f}_n - \widetilde{g}_n||_1.$$

Since $\mathcal{H}_1$ and $\mathcal{H}_2$ are convex and closed in measure, $f \in \mathcal{H}_1$ and $g \in \mathcal{H}_2$. By Fatou's lemma we get

$$m \leq ||f - g||_1 \leq \liminf_n ||\widetilde{f}_n - \widetilde{g}_n||_1 = m.$$ □

6. Convergences in $L^1_X(\mu)$

Now we proceed to the characterizations of several modes of convergences from a bounded subset in $L^1_X(\mu)$ via weak compactness of the truncated sequences.

Theorem 6.1. *Let X be a Banach space and $\mathcal{H}$ a bounded subset of $L^1_X(\mu)$. Then the following are equivalent:*

(1) *Given any sequence (f_n) in $\mathcal{H}$, there exists a sequence (g_n) with $g_n \in co\{f_k : k \geq n\}$ such that (g_n) converges a.e in X.*
(2) *Given any sequence (f_n) in $\mathcal{H}$, there exists a sequence (g_n) with $g_n \in co\{f_k : k \geq n\}$ such that (g_n) weakly converges a.e in X (i.e $\mathcal{H}$ has the WTP).*
(3) *Given any sequence (f_n) in $\mathcal{H}$, there exists a subsequence (g_n) of (f_n) such that $(1_{\{\|h_n\|<n\}}h_n)$ is $\sigma(L^1, L^\infty)$ relatively compact for each subsequence (h_n) of (g_n).*
(4) *Given any sequence (f_n) in $\mathcal{H}$, there exists a subsequence (g_n) of (f_n) such that $(1_{\{\|g_n\|<n\}}g_n)$ is $\sigma(L^1, L^\infty)$ relatively compact.*
(5) *Given any sequence (f_n) in $\mathcal{H}$, there exists a subsequence (f_{n_k}) of (f_n) such that $(1_{\{\|f_{n_k}\|<n_k\}}f_{n_k})$ converges $\sigma(L^1, L^\infty)$.(*)*
(6) *Given any sequence (f_n) in $\mathcal{H}$, there exists a subsequence (f_{n_k}) of (f_n) such that $f_{n_k} = u_{n_k} + v_{n_k}$, where (u_{n_k}) converges $\sigma(L^1, L^\infty)$ and (v_{n_k}) converges a.e to 0.*

Proof. (1)$\Longrightarrow$(2) is obvious. (2) $\Longrightarrow$ (3). Let (f_n) be a sequence in $\mathcal{H}$. By Lemma 4.1(a) there is a subsequence (g_n) of (f_n) such that $(1_{\{\|h_n\|<n\}}h_n)$ is uniformly integrable, for each subsequence (h_n) of (g_n). If $(1_{\{\|h_{n_p}\|<n_p\}}h_{n_p})$ is a subsequence of $(1_{\{\|h_n\|<n\}}h_n)$ there is, by (2), a sequence (u_p) with $u_p \in co\{h_{n_k} : k \geq p\}$ such that (u_p) weakly converges a.e in X. We have

$$u_p = \sum_{i=p}^{k_p} \lambda_i^p h_{n_i} = \sum_{i=p}^{k_p} \lambda_i^p 1_{\{\|h_{n_i}\|<n_i\}}h_{n_i} + \sum_{i=p}^{k_p} \lambda_i^p (h_{n_i} - 1_{\{\|h_{n_i}\|<n_i\}}h_{n_i})$$

with $0 \leq \lambda_i^p \leq 1$ and $\sum_{i=p}^{k_p} \lambda_i^p = 1$. By Lemma 4.1 (b), we have

$$\sum_{i=p}^{k_p} \lambda_i^p (h_{n_i} - 1_{\{\|h_{n_i}\|<n_i\}}h_{n_i}) \to 0$$

a.e as $p \to \infty$. Hence $\sum_{i=p}^{k_p} \lambda_i^p 1_{\{\|h_{n_i}\|<n_i\}}h_{n_i}$ weakly converges a.e as $p \to \infty$. By Theorem 3.3 we conclude that $(1_{\{\|h_n\|<n\}}h_n)$ is $\sigma(L^1, L^\infty)$ relatively compact. (3) $\Longrightarrow$ (4) is obvious. Let us prove that (4)$\Longrightarrow$(5). By Eberlein-Smulian

(*) Consequently $(f^k_{n_k}) = (1_{\{\|f_{n_k}\|<k\}}f_{n_k})$ converges $\sigma(L^1, L^\infty)$ too.

theorem there is a subsequence $(f_{n_k})_k$ of (f_n) such that $(f_{n_k}^{n_k})_k$ converges $\sigma(L^1, L^\infty)$ and so is the sequence $(f_{n_k}^k)_k$. Indeed we have

$$f_{n_k}^{n_k} - f_{n_k}^k = 1_{\{k \le ||f_{n_k}|| < n_k\}} f_{n_k} = 1_{\{k \le ||f_{n_k}^{n_k}||\}} f_{n_k}^{n_k}.$$

Since $(f_{n_k}^{n_k})_k$ is uniformly integrable, $f_{n_k}^{n_k} - f_{n_k}^k \to 0$ strongly in $L_X^1(\mu)$ as $k \to \infty$. It follows that $(f_{n_k}^k)_k$ converges $\sigma(L^1, L^\infty)$. (5)$\Longrightarrow$(6). Let (f_n) be a sequence in $\mathcal{H}$. By (5) there exists a subsequence (f_{n_k}) of (f_n) such that $(f_{n_k}^{n_k})_k$ converges $\sigma(L^1, L^\infty)$ in $L_X^1(\mu)$. Put $u_{n_k} = f_{n_k}^{n_k}$ and $v_{n_k} = f_{n_k} - u_{n_k}$. Then by Lemma 4.1 (5)$\Longrightarrow$(6) follows immediately.

(6)$\Longrightarrow$(1) Let us apply the notations in (6). Since (u_{n_k}) converges $\sigma(L^1, L^\infty)$ in $L_X^1(\mu)$, there is a sequence $(\widetilde{u}_m)$ of the form $\widetilde{u}_m = \sum_{i=m}^{\nu_m} \lambda_i^m u_{n_i}$ with $\lambda_i^m \ge 0$, $\sum_{i=m}^{\nu_m} \lambda_i^m = 1$ such that $(\widetilde{u}_m)$ converges strongly in $L_X^1(\mu)$. There is a subsequence $(\widetilde{u}_{m_p})$ which converges a.e in X and so is the sequence $(\sum_{i=m_p}^{\nu_{m_p}} \lambda_i^{m_p} f_{n_i})_p$. □

Combining Theorem 6.1 and Eberlein-Smulian theorem allows us to recover a criteria of weak compactness [23, 35] in $L_X^1(\mu)$:

Proposition 6.2. *Let X be a Banach space and $\mathcal{H}$ a bounded subset of $L_X^1(\mu)$. Then the following conditions are equivalent :*

(1) *$\mathcal{H}$ is $\sigma(L^1, L^\infty)$ relatively compact.*
(2) *$\mathcal{H}$ is uniformly integrable and has any one of the equivalent properties of Theorem 6.1.*

Comments. The equivalence (2)$\Longleftrightarrow$(6) in Theorem 6.1 was stated by a different method in [9, 10] via biting lemma (see, for example, [25, 28, 33]) whereas (2)$\Longleftrightarrow$(4)$\Longleftrightarrow$(5) characterize the WTP sets by weak compactness of the truncated sequences from bounded sequences in these sets. We would like to mention that Lemma 4.1 provides also an easy ingredient in several proofs of the results we present here.

In the following we show that WCP$\Longrightarrow$WTP.

Proposition 6.3. *Let X be a Banach space and $\mathcal{H}$ a subset of $L_X^1(\mu)$. If $\mathcal{H}$ has the WCP, then $\mathcal{H}$ has the WTP.*

Proof. Let (f_n) be a bounded sequence in $\mathcal{H}$. Since $\mathcal{H}$ has the WCP, there is a subsequence (h_n) of (f_n) such that $(\frac{1}{n}\sum_{i=1}^n h_i)$ weakly converges a.e in X. By Lemma 4.1, by extracting a subsequence if necessary, we may suppose that $h_n - 1_{\{||h_n|| < n\}} h_n \to 0$ a.e. Let us consider the sequence (u_n) defined by

$$u_n = \frac{1}{n^3}(n h_n + \sum_{i=n+1}^{n^3} h_i).$$

Then

$$u_n \in co\{h_k : k \geq n\} \subset co\{f_k : k \geq n\}$$

and

$$\begin{aligned} u_n &= \frac{1}{n^3}\sum_{i=1}^{n^3} h_i + \frac{1}{n^2}h_n - \frac{1}{n^3}\sum_{i=1}^{n} h_i \\ &= \frac{1}{n^3}\sum_{i=1}^{n^3} h_i + \frac{1}{n^2}1_{\{||h_n||<n\}}h_n \\ &+ \frac{1}{n^2}(h_n - 1_{\{||h_n||<n\}}h_n) - \frac{1}{n^2}(\frac{1}{n}\sum_{i=1}^{n} h_i). \end{aligned}$$

As $\frac{1}{n^3}\sum_{i=1}^{n^3} h_i$ weakly converges a.e, $\frac{1}{n^2}1_{\{||h_n||<n\}}h_n$ strongly converges a.e to 0, $\frac{1}{n^2}(h_n - 1_{\{||h_n||<n\}}h_n)$ strongly converges a.e to 0 and $\frac{1}{n^2}(\frac{1}{n}\sum_{i=1}^{n} h_i)$ strongly converges a.e to 0 because $\frac{1}{n}\sum_{i=1}^{n} h_i$ weakly converges a.e, (u_n) weakly converges a.e thus showing that $\mathcal{H}$ has the WTP. □

It is worthy to address the question of the equivalence WTP$\Longleftrightarrow$WKP. For this we need a technical lemma via Komlós theorem and Theorem 6.1 . See [1, 3, 4, 5, 6, 7, 19, 31, 32] for related results.

Lemma 6.4. *Suppose that X is a separable Banach space, $(e^*_p)_{p\in\mathbb{N}}$ is a sequence in X^* and (u_n) is a bounded WTP sequence in $L^1_X(\mu)$, then there exist a subsequence (v_m) , $\varphi \in L^1_{\mathbb{R}^+}(\mu)$ and $u \in L^1_X(\mu)$ such that for all $p \in \mathbb{N}$ and almost all $\omega \in \Omega$*

$$\lim_{n\to\infty} \frac{1}{n}\Sigma^n_{j=1}||w_j(\omega)|| = \varphi(\omega)$$

and

$$\lim_{n\to\infty} \langle e^*_p, \frac{1}{n}\Sigma^n_{j=1}w_j(\omega)\rangle = \langle e^*_p, u(\omega)\rangle$$

for each subsequence $(w_l) = (v_{m_l})$.

Proof. Since (u_n) is bounded WTP, by Theorem 6.1 ((2)$\Longrightarrow$ (6)), there is a subsequence (u'_n) with $u'_n = v'_n + w'_n$ where (v'_n) converges $\sigma(L^1, L^\infty)$ to u in $L^1_X(\mu)$ and $||w'_n|| \to 0$ a.e. Thus we may suppose that (u_n) converges $\sigma(L^1, L^\infty)$ to u. Using Komlós theorem [29] and an appropriate diagonal procedure, we find a subsequence $(v_m) = (u_{n_m})$, $\varphi \in L^1_{\mathbb{R}^+}(\mu)$ and $(\varphi_p)_p$ in $L^1_{\mathbb{R}}(\mu)$ such that for all $p \in \mathbb{N}$ and almost all $\omega \in \Omega$, the following hold:

$$\lim_{n\to\infty} \frac{1}{n}\Sigma^n_{j=1}||v_{m_j}(\omega)|| = \varphi(\omega)$$

$$\lim_{n\to\infty} \frac{1}{n}\Sigma^n_{j=1}\langle e^*_p, v_{m_j}(\omega)\rangle = \varphi_p(\omega)$$

for each subsequence $(w_l) = (v_{m_l})$. Since (u_n) converges $\sigma(L^1, L^\infty)$ to u, so is (v_{m_l}). It follows that

$$\forall p \in \mathbb{N}, \ \frac{1}{n}\Sigma_{j=1}^n \langle e_p^*, v_{m_j}\rangle \overset{\sigma(L^1,L^\infty)}{\rightarrow} \langle e_p^*, u\rangle.$$

Whence we get

$$\forall p \in \mathbb{N}, \ \varphi_p = \langle e_p^*, u\rangle,$$

thus completing the proof. □

Remarks. Let us mention some particular cases of Lemma 6.4.

(1) If X is reflexive, $L^1_X(\mu)$ has the WTP (see, for example, [9, 10]).

(2) If (u_n) is a bounded sequence in $L^1_X(\mu)$ and $\bigcup_n\{u_n(\omega)\}$ is ball-relatively weakly compact for every $\omega \in \Omega$, that is its intersection with any closed ball is relatively weakly compact, then (u_n) has the WTP.

In both these cases, we can also apply Lemma 4.2 (instead of Theorem 6.1) which provides the $\sigma(L^1, L^\infty)$ convergence of the truncated sequences into consideration.

Now we are able to state the equivalence WTP$\Longleftrightarrow$WKP.

Proposition 6.5. *Suppose that X is a separable Banach space, $\Gamma : \Omega \to X$ is a $\mathcal{R}_w(X)$-valued measurable multifunction. Then the set $\mathcal{S}^1_\Gamma$ of all integrable selections of Γ has the WKP.*

Proof. By hypothesis $\mathcal{S}^1_\Gamma$ is obviously $\mathcal{R}_w(X)$-tight. By [9,10] we know that $\mathcal{R}_w(X)$-tight$\Longrightarrow$WTP. Let (u_n) be a bounded sequence in $\mathcal{S}^1_\Gamma$. Let us consider a dense sequence $(e_p^*)_{p\in\mathbb{N}}$ for the Mackey topology $\tau(X^*, X)$. As (u_n) is bounded WTP, using Lemma 6.4 provides a subsequence (v_m) , $\varphi \in L^1_{\mathbb{R}^+}(\mu)$ and $u \in L^1_X(\mu)$ such that for all $p \in \mathbb{N}$ and almost all $\omega \in \Omega$

$$\lim_{n\to\infty} \frac{1}{n}\Sigma_{j=1}^n ||w_j(\omega)|| = \varphi(\omega) \tag{6.5.1}$$

and

$$\lim_{n\to\infty} \langle e_p^*, \frac{1}{n}\Sigma_{j=1}^n w_j(\omega)\rangle = \langle e_p^*, u(\omega)\rangle \tag{6.5.2}$$

for each subsequence $(w_l) = (v_{m_l})$. By (6.5.1) we see that $(\frac{1}{n}\Sigma_{j=1}^n w_j(.))_n$ is pointwise bounded and by hypothesis $\frac{1}{n}\Sigma_{j=1}^n w_j(\omega) \in \Gamma(\omega)$ for almost all $\omega \in \Omega$ and for all $n \in \mathbb{N}^*$. Since Γ is $\mathcal{R}_w(X)$-valued, it follows that

$$\frac{1}{n}\Sigma_{j=1}^n w_j(\omega) \in \Delta(\omega)$$

for almost all ω where Δ is a measurable *convex weakly compact* valued multifunction (we can take $\Delta(\omega) = \Gamma(\omega)\bigcap(\varphi(\omega) + m(\omega))\overline{B}_E)$ where

$$m(.) := \sup_n \Sigma_{j=1}^n ||w_j(.)||.$$

Using the density of $(e^*_p)_{p\in\mathbb{N}}$, and (6.5.2) shows that, for almost all $\omega \in \Omega$ and for all $e^* \in X^*$,

$$\lim_{n\to\infty} \langle e^*, \frac{1}{n}\Sigma_{j=1}^n w_j(\omega)\rangle = \langle e^*, u(\omega)\rangle \tag{6.5.3}$$

for each subsequence thus showing that (u_n) has the WKP. □

Remarks. 1. If $cwk(X)$ denotes the collection of non empty convex weakly compact subsets of X, Saadoune [31, Theorem 1] introduced the following tighness condition, namely for every sequence (u_n) in the set into consideration, there is a sequence $(\widetilde{u}_n)$ satisfying: (i) $\widetilde{u}_n \in co\{u_m : m \geq n\}$ for every n, (ii) for every $\varepsilon > 0$, there is a convex $cwk(X)$-valued measurable multifunction $\Gamma_\varepsilon : \Omega \to X$ such that

$$\sup_n \mu(\{\omega \in \Omega : \widetilde{u}_n(\omega) \notin \Gamma_\varepsilon(\omega)\}) \leq \varepsilon.$$

Actually $cwk(X)$-tight $\Longrightarrow$ $\mathcal{R}_w(X)$-tight $\Longrightarrow$ WTP. An inspection of the proofs of Lemma 6.4, Proposition 6.5 and [31, Theorem 1] shows that the minimal tightness assumption is the following generalization of the WTP for subsets in $L^1_X(\mu)$. For every sequence (u_n) in the set into consideration, there is a $\mathcal{R}_w(X)$-tight sequence $(\widetilde{u}_n)$ such that $\widetilde{u}_n \in co\{u_m : m \geq n\}$ for every n. An easy example of unbounded $\mathcal{R}_w(X)$-tight set is $L^1_X(\mu)$ where X is a separable reflexive Banach space. It is obvious that the properties "$cwk(X)$-tight" and "$\mathcal{R}_w(X)$-tight"'are equivalent on bounded sequences of $L^1_X(\mu)$.

2. If the sequence (u_n) is bounded $cwk(X)$-tight in $L^1_X(\mu)$, then there exists $u \in L^1_X(\mu)$, such that $\forall h \in L^\infty_{X^*}(\mu), (\langle h, u_n\rangle)$ Komlós converges in measure to $\langle h, u\rangle$ [31, Theorem 1].

Let us mention the following variant. See also Díaz [24].

Proposition 6.6. *Suppose that X is a Banach space with strongly separable dual, and $\mathcal{H}$ is a WTP subset of $L^1_X(\mu)$. Then $\mathcal{H}$ has the WKP.*

Proof. Let $D^* := (e^*_p)_{p\in\mathbb{N}}$ be a dense sequence in X^* for the norm topology. Let (f_n) be a bounded sequence in $\mathcal{H}$. By Lemma 6.4, there exist a subsequence (g_m) of (f_n), $\zeta \in L^1_{\mathbb{R}^+}(\mu)$ and $f \in L^1_X(\mu)$ such that for all $p \in \mathbb{N}$ and almost all $\omega \in \Omega$

$$\lim_{n\to\infty} \frac{1}{n}\Sigma_{j=1}^n ||h_j(\omega)|| = \zeta(\omega) \tag{6.6.1}$$

and

$$\lim_{n\to\infty} \langle e^*_p, \frac{1}{n}\Sigma_{j=1}^n h_j(\omega)\rangle = \langle e^*_p, f(\omega)\rangle \tag{6.6.2}$$

for each subsequence $(h_l) = (g_{m_l})$. By (6.6.1) $(\frac{1}{n}\Sigma_{j=1}^n h_j)_n$ is pointwise bounded so that by (6.6.2) and the density of D^*, we get, for almost all $\omega \in \Omega$, and for all $e^* \in X^*$,

$$\lim_{n\to\infty} \langle e^*, \frac{1}{n}\Sigma_{j=1}^n h_j(\omega)\rangle = \langle e^*, f(\omega)\rangle$$

for each subsequence. This proves that $\mathcal{H}$ has the WKP.

Comments. 1. Apart from the use of Lemma 4.1, 4.2 and Theorem 6.1, Komlós arguments are not new since they have been employed in several places by Balder [3, 4, 5, 6] and Balder-Hess [7]. Recent applications of Komlós theorem are given by [1, 7, 31, 32] via tighness conditions.

2. If X is B-convex and reflexive, any bounded sequence in $L^1_X(\mu)$ has the strong Komlós property (SKP) by virtue of a result due to Bourgain [12]. Applications of vector-valued Komlós theorem due to Bourgain [12] and Garling [26] to Komlós-Mosco convergence for convex weakly compact random sets in B-convex reflexive separable Banach spaces are given in [17, 19]. An elementary proof of the SKP in $L^1_X(\mu)$ where X is an Hilbert space is given in Guessous [27].

3. Propositions 6.5 and 6.6 provide the characterizations of weak compactness in $L^1_X(\mu)$ in terms of weak Komlós convergence. See [24, 31, 32] for details.

4. The characterizations of weak compactness and weak conditionally compactness in the space $L^1_{E'}[E]$ (E being a Banach space) of scalarly integrable E'-valued functions via WTP appear in a forthcoming paper by Benabdellah-Castaing [11].

It is worthy to pose the problem of characterizing the Banach X by the previous modes of convergences. The following result gives a positive answer to this question.

Proposition 6.7. *Let X be a Banach space. Then the following are equivalent:*

(1) *$L^1_X(\mu)$ has the WKP.*
(2) *$L^1_X(\mu)$ has the WCP.*
(3) *$L^1_X(\mu)$ has the WTP.*
(4) *X is reflexive.*

Proof. (1)$\Longrightarrow$(2) is obvious. (2)$\Longrightarrow$(3) follows from Proposition 6.3. (3)$\Longrightarrow$(4). Let (x_n) be a bounded sequence in X and $M = \sup_n ||x_n||$. Put $\mathcal{H} = \{1_\Omega x_n : n \geq 1\}$. Then $\mathcal{H}$ is a bounded subset of $L^1_X(\mu)$. By (3) $\mathcal{H}$ has the WTP, so by Theorem 6.1 ((2) $\Longrightarrow$ (3) $\Longrightarrow$ (5) and footnote (*)), there is a subsequence $(1_\Omega x_{n_k})_k$ of $(1_\Omega x_n)$ such that $(1_\Omega x^k_{n_k})_k$ is $\sigma(L^1, L^\infty)$ convergent. Remark that if $k > M$, then $1_\Omega x^k_{n_k} = 1_\Omega x_{n_k}$. So $(1_\Omega x_{n_k})_k$ is $\sigma(L^1, L^\infty)$ convergent and therefore $(x_{n_k})_k$ is $\sigma(X, X^*)$ convergent. Whence X is reflexive. (4)$\Longrightarrow$(1). We can suppose that X is separable. It is enough to apply Proposition 6.5 to the constant multifunction $\Gamma(.) = X$. □

7. Conditionally WTP sets in $L^1_X(\mu)$

In [9, 10], the authors characterizes the conditionally weakly compact subsets of $L^1_X(\mu)$ (cf. Theorem 3.4) as bounded uniformly integrable subsets satisfying:

(**) given any sequence (u_n) in the set into consideration, there exists a sequence (v_n), with $v_n \in co\{u_m : m \geq n\}$ such that $(v_n(\omega))$ is weakly Cauchy in X for almost all $\omega \in \Omega$.

Now taking into account the Proposition 6.7 it is worthy to pose the question of characterizing Banach spaces X not containing l^1 by the foregoing "conditionally weak compactness condition". We will give a positive answer to this question that leads to the unsual modes of convergence we introduce below. We will need several analoguous versions of the results developped in Section 6. The first one is the analog of Theorem 6.1 which based on Theorem 3.4 and the arguments of the proof of Theorem 6.1.

Theorem 7.1. *Let $\mathcal{H}$ be a bounded subset of $L^1_X(\mu)$. Then the following are equivalent:*

(1) *Given any sequence (f_n) in $\mathcal{H}$, there exists a sequence (g_n) with $g_n \in co\{f_k : k \geq n\}$ such that (g_n) is weakly Cauchy a.e in X.*
(2) *Given any sequence (f_n) in $\mathcal{H}$, there exists a subsequence (g_n) of (f_n) such that $g_n = u_n + v_n$ where (u_n) is weakly Cauchy in $L^1_X(\mu)$ and $||v_n|| \to 0$ a.e.*
(3) *Given any sequence (f_n) in $\mathcal{H}$, there exists a subsequence (g_n) of (f_n) such that $(1_{\{||h_n||<n\}}h_n)$ is conditionally weakly compact for each subsequence (h_n) of (g_n).*
(4) *Given any sequence (f_n) in $\mathcal{H}$, there exists a subsequence (g_n) of (f_n) such that $(1_{\{||g_n||<n\}}g_n)$ is conditionally weakly compact.*
(5) *Given any sequence (f_n) in $\mathcal{H}$, there exists a subsequence (g_n) of (f_n) such that $(1_{\{||g_n||<n\}}g_n)$ is weakly Cauchy in $L^1_X(\mu)$.*

Proof. (1)$\Longrightarrow$(3). Let (f_n) be a sequence in $\mathcal{H}$. By Lemma 4.1(a) there is a subsequence (g_n) of (f_n) such that $(1_{\{||h_n||<n\}}h_n)$ is uniformly integrable, for each subsequence (h_n) of (g_n). If $(1_{\{||h_{n_p}||<n_p\}}h_{n_p})$ is a subsequence of $(1_{\{||h_n||<n\}}h_n)$ there is, by (1), a sequence (u_p) with $u_p \in co\{h_{n_k} : k \geq p\}$ such that (u_p) is weakly Cauchy a.e in X. We have

$$u_p = \sum_{i=p}^{k_p} \lambda_i^p h_{n_i} = \sum_{i=p}^{k_p} \lambda_i^p 1_{\{||h_{n_i}||<n_i\}}h_{n_i} + \sum_{i=p}^{k_p} \lambda_i^p (h_{n_i} - 1_{\{||h_{n_i}||<n_i\}}h_{n_i})$$

with $0 \le \lambda_i^p \le 1$ and $\sum_{i=p}^{k_p} \lambda_i^p = 1$. By Lemma 4.1 (b), we have

$$\sum_{i=p}^{k_p} \lambda_i^p (h_{n_i} - 1_{\{||h_{n_i}||<n_i\}} h_{n_i}) \to 0$$

a.e as $p \to \infty$. Hence $(v_p) = (\sum_{i=p}^{k_p} \lambda_i^p 1_{\{||h_{n_i}||<n_i\}} h_{n_i})_p$ is weakly Cauchy a.e. By Theorem 3.4 we conclude that (h_n^n) is conditionally weakly compact. (3) $\Longrightarrow$ (4) is obvious. Let us prove that (4)$\Longrightarrow$(5). Suppose that $(g_n^n)_n$ is conditionally weakly compact and $(g_{n_k}^{n_k})_k$ is a weakly Cauchy subsequence in $L_X^1(\mu)$. Then we have

$$g_{n_k}^{n_k} - g_{n_k}^k = 1_{\{k \le ||g_{n_k}|| < n_k\}} g_{n_k} = 1_{\{k \le ||g_{n_k}^{n_k}||\}} g_{n_k}^{n_k}.$$

Since $(g_{n_k}^{n_k})_k$ is uniformly integrable, $g_{n_k}^{n_k} - g_{n_k}^k \to 0$ strongly in $L_X^1(\mu)$ as $k \to \infty$. It follows that $(g_{n_k}^k)_k$ is weakly Cauchy in $L_X^1(\mu)$. (5)$\Longrightarrow$(2). Let (f_n) be a sequence in $\mathcal{H}$ and let (g_n) be a sequence given by Lemma 4.1. By (5) there is a subsequence (h_n) of (g_n) such that (h_n^n) is weakly Cauchy in $L_X^1(\mu)$. Put $f_n' = h_n, u_n = h_n^n$ and $v_n = h_n - h_n^n$, then (5)$\Longrightarrow$(2) follows.

(2)$\Longrightarrow$(1) Since (u_n) is weakly Cauchy in $L_X^1(\mu)$, it is conditionally weakly compact. By Theorem 3.4 there is a sequence $(\widetilde{u}_n)$ of the form $\widetilde{u}_n = \sum_{i=n}^{\nu_n} \lambda_i^n u_i$ with $\lambda_i^n \ge 0$, $\sum_{i=n}^{\nu_n} \lambda_i^n = 1$ such that $(\widetilde{u}_n)$ is weakly Cauchy a.e. Hence the sequence $(g_n) = (\sum_{i=n}^{\nu_n} \lambda_i^n f_i)_n$ is weakly Cauchy a.e. □

We introduce the following definitions. Let $\mathcal{H}$ be a subset in $L_X^1(\mu)$.

(1) $\mathcal{H}$ is said to have the *conditionally weak Talagrand property* (CWTP) if, given any bounded sequence (f_n) in $\mathcal{H}$, there exists a sequence (g_n) with $g_n \in co\{f_k : k \ge n\}$ such that (g_n) is weakly Cauchy a.e in X.
(2) $\mathcal{H}$ is said to have the *conditionally weak Césaro property* (CWCP) if, given any bounded sequence (f_n) in $\mathcal{H}$, there exists a subsequence (g_n) such that the Césaro sums $(\frac{1}{n}\Sigma_{i=1}^n g_i)_n$ is weakly Cauchy a.e in X.
(3) $\mathcal{H}$ is said to have the *conditionally weak Komlós property* (CWKP) if, given any bounded sequence (f_n) in $\mathcal{H}$, there exist a subsequence (g_n) such that $(\frac{1}{n}\Sigma_{i=1}^n g_i)_n$ is weakly Cauchy in X a.e and $\frac{1}{n}\Sigma_{i=1}^n g_i - \frac{1}{n}\Sigma_{i=1}^n h_i \to 0$ weakly a.e. for each subsequence (h_n) of (g_n).

In the following we show that CWCP$\Longrightarrow$CWTP.

Proposition 7.2. *Let X be a Banach space and $\mathcal{H}$ a subset of $L_X^1(\mu)$. If $\mathcal{H}$ has the CWCP, then $\mathcal{H}$ has the CWTP.*

Proof. Let (f_n) be a bounded sequence in $\mathcal{H}$. By Lemma 4.1, there is a subsequence (g_n) of (f_n) such that $h_n - 1_{\{||h_n||<n\}} h_n \to 0$ a.e for each subsequence (h_n) of (g_n). Since $\mathcal{H}$ has the CWCP, there a subsequence (h_n) of (g_n) such that $(\frac{1}{n}\sum_{i=1}^n h_i)$ is weakly Cauchy a.e in X. Let us consider the sequence (u_n) defined by

$$u_n = \frac{1}{n^3}(nh_n + \sum_{i=n+1}^{n^3} h_i).$$

Then

$$u_n \in co\{h_k : k \geq n\} \subset co\{f_k : k \geq n\}$$

and

$$\begin{aligned} u_n &= \frac{1}{n^3}\sum_{i=1}^{n^3} h_i + \frac{1}{n^2}h_n - \frac{1}{n^3}\sum_{i=1}^{n} h_i \\ &= \frac{1}{n^3}\sum_{i=1}^{n^3} h_i + \frac{1}{n^2}1_{\{||h_n||<n\}}h_n \\ &+ \frac{1}{n^2}(h_n - 1_{\{||h_n||<n\}}h_n) - \frac{1}{n^2}(\frac{1}{n}\sum_{i=1}^{n} h_i). \end{aligned}$$

As $\frac{1}{n^3}\sum_{i=1}^{n^3} h_i$ is weakly Cauchy a.e, $\frac{1}{n^2}1_{\{||h_n||<n\}}h_n$ strongly converges a.e to 0, $\frac{1}{n^2}(h_n - 1_{\{||h_n||<n\}}h_n)$ strongly converges a.e to 0 and $\frac{1}{n^2}(\frac{1}{n}\sum_{i=1}^{n} h_i)$ strongly converges a.e to 0 because $\frac{1}{n}\sum_{i=1}^{n} h_i$ is weakly Cauchy a.e, (u_n) is weakly Cauchy a.e, showing that $\mathcal{H}$ has the CWTP. □

There is an easy lemma.

Lemma 7.3. *Let (f_n) be a sequence in $L^1_X(\mu)$. If (f_n) weakly converges to $f \in L^1_X(\mu)$ and if (f_n) is weakly Cauchy a.e, then (f_n) weakly converges a.e to f.*

Proposition 7.4. *Let $\mathcal{H}$ be a subset of $L^1_X(\mu)$.*

(a) *If $\mathcal{H}$ has both the WTP and CWCP, then $\mathcal{H}$ has the WCP.*

(b) *If $\mathcal{H}$ has both the WTP and CWKP, then $\mathcal{H}$ has the WKP.*

Proof. Let (f_n) be a bounded sequence in $\mathcal{H}$. Since $\mathcal{H}$ has the WTP, there exist a subsequence (g_n) of (f_n) such that $g_n = u_n + v_n$ where (u_n) weakly converges to $u \in L^1_X(\mu)$ and $||v_n|| \to 0$ a.e.

(a) Since $\mathcal{H}$ has the CWCP, there is a subsequence $(g'_n) = (u'_n + v'_n)$ such that $(\frac{1}{n}\Sigma_{i=1}^n g'_i)$ is weakly Cauchy a.e, then $(\frac{1}{n}\Sigma_{i=1}^n u'_i)$ weakly converges to u in $L^1_X(\mu)$ and is weakly Cauchy a.e. By Lemma 7.3 $(\frac{1}{n}\Sigma_{i=1}^n u'_i)$ weakly converges a.e to u; so is $(\frac{1}{n}\Sigma_{i=1}^n g'_i)$. Hence $\mathcal{H}$ has the WCP.

(b) Since $\mathcal{H}$ has the CWKP, there is a subsequence (g'_n) of (g_n) such that

$$\lim_{n\to\infty}[\frac{1}{n}\Sigma_{i=1}^n g'_i - \frac{1}{n}\Sigma_{i=1}^n h'_i] = 0$$

weakly a.e for each subsequence (h'_n) of (g'_n). Therefore $\mathcal{H}$ has the WKP. □

The following result is useful in the characterization of the CWKP in $L^1_X(\mu)$.

Lemma 7.5. *Let X be a Banach space, (x_n) a bounded sequence in X satisfying:*

(i) *there is a sequence $D^* := (e^*_p)$ in X^* which separates the points of*

$$\overline{co\{x_n : n \in \mathbb{N}^*\}}^{\sigma(X^{**},X^*)}$$

where $\overline{A}^{\sigma(X^{**},X^*)}$ denotes the $\sigma(X^{**},X^*)$ closure of A,
(ii) *(x_n) is $\sigma(X, D^*)$-Cauchy,*
(iii) *$\{x_n : n \in \mathbb{N}^*\}$ is relatively sequentially $\sigma(X^{**},X^*)$ compact.*

Then (x_n) is weakly Cauchy in X.

Proof. Let $x^* \in X^*$ and let us denote by α and β two cluster points of $(\langle x^*, x_n\rangle)_n$. There are subsequences (u_n) and (v_n) such that

$$\langle x^*, u_n\rangle \to \alpha \text{ and } \langle x^*, v_n\rangle \to \beta.$$

By (iii) there are subsequences (u'_n) and (v'_n), u and v in X^{**} such that

$$u'_n \to u \text{ and } v'_n \to v$$

for the $\sigma(X^{**},X^*)$ topology. By (ii) it follows that

$$\langle e^*_p, u\rangle = \lim_n \langle e^*_p, u'_n\rangle = \lim_n \langle e^*_p, x_n\rangle.$$

and

$$\langle e^*_p, v\rangle = \lim_n \langle e^*_p, v'_n\rangle = \lim_n \langle e^*_p, x_n\rangle.$$

for every p. As (e^*_p) separates the points of $\overline{co\{x_n : n \in \mathbb{N}^*\}}^{\sigma(X^{**},X^*)}$, it follows that $u = v$. But

$$\alpha = \lim_n \langle x^*, u_n\rangle = \lim_n \langle x^*, u'_n\rangle = \langle x^*, u\rangle$$

and

$$\beta = \lim_n \langle x^*, v_n\rangle = \lim_n \langle x^*, v'_n\rangle = \langle x^*, v\rangle.$$

Therefore $\alpha = \beta$. □

Remarks. If X has no copy of l^1, condition (iii) is satisfied, because the closed unit ball $\overline{B}_{X^{**}}$ is sequentially $\sigma(X^{**},X^*)$ compact by the Odell-Rosenthal theorem [22, p. 236]. This leads us to the following parametric version of Lemma 7.5.

Proposition 7.6. *Suppose that X is a Banach space with separable dual. Then $L^1_X(\mu)$ has the CWKP.*

Proof. Step 1. Claim: $L^1_X(\mu)$ has the CWTP. Since X^* is separable there is a dense sequence (e^*_p) in X^* which separates the points of X^{**} and X has no copy of l^1. Let (f_n) be a bounded sequence in $L^1_X(\mu)$. Using Komlós theorem [29] and an appropriate diagonal procedure, we find a subsequence $(v_m) = (f_{n_m})$, $\varphi \in L^1_{\mathbb{R}+}(\mu)$ and $(\varphi_p)_p$ in $L^1_{\mathbb{R}}(\mu)$ such that for all $p \in \mathbb{N}$ and almost all $\omega \in \Omega$, the following hold:

$$\lim_{n\to\infty} \frac{1}{n}\Sigma^n_{j=1}||w_j|| = \varphi \tag{7.6.1}$$

$$\lim_{n\to\infty} \frac{1}{n}\Sigma^n_{j=1}\langle e^*_p, w_j\rangle = \varphi_p \tag{7.6.2}$$

for each subsequence $(w_l) = (v_{m_l})$. By (7.6.1) the sequence $(\frac{1}{n}\Sigma^n_{j=1} v_j)_n$ is pointwise bounded a.e. so that by (7.6.2) and the above remark, we can apply Lemma 7.5 to $(\frac{1}{n}\Sigma^n_{j=1} v_j)_n$ showing that $(\frac{1}{n}\Sigma^n_{j=1} v_j)_n$ is weakly Cauchy a.e. Hence we conclude that $L^1_X(\mu)$ has the CWCP. By Proposition 7.2 (CWCP$\Longrightarrow$CWTP), the claim follows.

Step 2. Claim: $L^1_X(\mu)$ has the CWKP. Let (f_n) be a bounded sequence in $L^1_X(\mu)$. By Theorem 7.1 there is a subsequence (g_n) such that $g_n = x_n + y_n$ where (x_n) is weakly Cauchy in $L^1_X(\mu)$ and $||y_n|| \to 0$ a.e. Then we may suppose that (f_n) is weakly Cauchy in $L^1_X(\mu)$. From the definition of CWKP, it remains to prove that

$$\lim_n[\frac{1}{n}\Sigma^n_{j=1}v_j - \frac{1}{n}\Sigma^n_{j=1}w_j] = 0$$

weakly a.e for every subsequence (w_m) of (v_m) where (v_m) is the subsequence of (f_n) given in Step 1. With similar arguments as above, $(\frac{1}{n}\Sigma^n_{j=1}w_j)_n$ is weakly Cauchy a.e. For simplicity let us put $z_n := \frac{1}{n}\Sigma^n_{j=1}v_j - \frac{1}{n}\Sigma^n_{j=1}w_j$. Then (z_n) is weakly Cauchy a.e. Since (f_n) is weakly Cauchy in $L^1_X(\mu)$, $z_n \to 0$ weakly in $L^1_X(\mu)$. Applying Lemma 7.3 to (z_n) completes the proof. □

Corollary 7.7. *Suppose that X is a separable Banach space. Let us consider the following conditions:*

(1) *$L^1_X(\mu)$ has the CWKP.*

(2) *$L^1_X(\mu)$ has the CWCP.*

(3) *$L^1_X(\mu)$ has the CWTP.*

(4) *X has no copy of l^1.*

Then (1) $\Longrightarrow$ (2) $\Longrightarrow$ (3) $\Longrightarrow$ (4). *If X^* is separable, then* (4) $\Longrightarrow$ (1).

Proof. It is enough to prove (3) $\Longrightarrow$ (4) $\Longrightarrow$ (1) If (x_n) is a bounded sequence in X and $M := \sup_n ||x_n||$, the set $\mathcal{H} := \{1_\Omega x_n : n \in \mathbb{N}^*\}$ is bounded in $L^1_X(\mu)$. By Theorem 7.1 there is a subsequence $(1_\Omega x_{n_k})_k$ such that the sequence $(1_\Omega x^k_{n_k})_k$ is weakly Cauchy in $L^1_X(\mu)$. If $k > M$, $1_\Omega x^k_{n_k} = 1_\Omega x_{n_k}$, then $(1_\Omega x_{n_k})_k$ is weakly Cauchy in $L^1_X(\mu)$. Hence $(x_{n_k})_k$ is weakly Cauchy in X. So X has no copy of l^1. (4) $\Longrightarrow$ (1) follows from Proposition 7.6. □

Acknowledgement. We wish to thank M. Valadier for helpful suggestions and a careful reading of this paper.

References

[1] Amrani, A., Castaing C.: Weak compactness in Pettis integration. Bulletin Polish Acad. Sc. **45**, No2, 139-150 (1997)

[2] Amrani, A., Castaing, C., Valadier, M.: Méhodes de troncature appliquées à des problèmes de convergence faible ou forte dans L^1. Arch. Rational Mech. Anal. **117**, 167-191 (1992)

[3] Balder, E. J.: Infinite-dimensional extension of a theorem of Komlós. Probab. Theory Related fields **81**, 185-188 (1989)

[4] Balder, E. J.: New sequential compactness results for spaces of scarlarly integrable functions. J. M. A. A **151**, 1-16 (1990)

[5] Balder, E. J.: On Prohorov's theorem for transition probabilities. Sém. Anal. Convexe **19**, 9.1-9.11 (1989)

[6] Balder, E. J.: On equivalence of strong and weak convergence in L^1-spaces under extreme point conditions. Israel J. Math. **75**, 21-47 (1991)

[7] Balder, E. J., Hess C.: Two generalizations of Komlo's theorem with lower closure-type applications. Journal of Convex Analysis **3** (1), 25-44 (1996)

[8] Beer, G.: Topologies on closed and closed convex subsets and the Effros measurability of set valued functions. Sém. Anal. Convexe Montpellier **2**, 2.1-2.44 (1991)

[9] Benabdellah, H., Castaing, C.: Weak compactness and convergences in $L^1_E(\mu)$. C.R. Acad. Sci. Paris **321**, 165-170 (1995)

[10] Benabdellah, H., Castaing, C.: Weak compactness criteria and convergences in $L^1_E(\mu)$. Collectanea Mathematica **XLVIII**, 423-448 (1997)

[11] Benabdellah, H., Castaing, C.: Weak compactness and convergences in $L^1_{E'}[E]$. Université Montpellier II, 1996, Preprint 31 pages.

[12] Bourgain, J.: The Komlós theorem for vector valued functions. Wrije Universiteit Brussel (1979) (9 pages). Unpublished.

[13] Bukhvalov, A.V.: Optimization without compactness, and Its applications. In:Operator theory: Advances and Applications Vol 75, pp.95-112. Birkhäuser Verlag 1995

[14] Bukhvalov, A.V., Lozanovskii, G.Ya.: On sets closed with respect to convergence in measure in spaces of measurable functions. Dokl. Akad. Nauk SSSR **212**, 1273-1275 (1973); Englis transl. Soviet Math. Dokl. 1563-1565 (1973)

[15] Castaing, C.: Quelques résultats de convergence des suites adaptées. Sém. Anal. Convexe Montpellier **17**, 2.1-2.24 (1987)

[16] Castaing, C.: Méthodes de compacité et de décomposition, Applications : Minimisation, convergence des martingales, lemme de Fatou multivoque. Ann. Mat. Pura Appli. **164**, 51-75 (1993)

[17] Castaing, C.: Weak compactness and convergences in Bochner and Pettis integration. Vietnam Journal of Math. **24** (3), 241-286 (1996)
[18] Castaing, C., Clauzure, P.: Compacité faible dans l'espace L^1_E et dans l'espace des multifonctions intégrablement bornées et minimisation. Ann. Mat. Pura Appl. **4** (140), 345-364 (1985)
[19] Castaing, C., Ezzaki, F.: Convergences for convex weakly compact random sets in B-convex reflexive Banach spaces, Supplemento al Vol. XLVI, 123-149 (1998) Atti Sem. Mat. Univ. Modena
[20] Castaing, C., Valadier, M.: Convex Analysis and Measurable multifunctions. Lecture Notes in Mathematics **580**, Springer 1977
[21] Chatterji, S.D.: A subsequence principle in probability theory. Jber.d. Dt. Math.-Verein **87**, 91-107 (1985)
[22] Diestel, J.: Geometry of Banach Spaces, Selected topics. Lectures Notes in Mathematics **485**, Springer 1975
[23] Diestel, J., Ruess, W.M., Schachermeyer, W.: Weak compactness in $L^1(\mu, X)$. Proc. Amer. Math. Soc. **118** (2), 447-453 (1993)
[24] Díaz, S.: Weak compactness in $L^1(\mu, X)$, Proc. Amer. Math. Soc. **124** (9), 2685-2693 (1996)
[25] Gaposkhin, V. F.: Convergence and limit theorems for sequences of random variables. Theory Probab. Appl. **17**, 379-400 (1972)
[26] Garling, D.J.H.: Subsequence principles for vector-valued random variables. Math. Proc. Camb. Phil. Soc. **86**, 301-311 (1979)
[27] Guessous, M.: An elementary proof of Komlós-Revész theorem in Hilbert spaces. Journal of Convex Analysis **4**, 321-332 (1997)
[28] Kadec, M.I., Pelczynski, A.: Bases, lacunary sequences and complemented subspaces in the spaces L^p. Studia Math. **21**, 161-176 (1962)
[29] Komlós, J.: A generalisation of a problem of Steinhaus. Acta Math. Acad. Sci. Hungar. **18**, 217-229 (1967)
[30] Levin, V.L.: Extremal problems with convex functionals that are lower semicontinuous with respect to convergence in measure. Dokl. Akad. Nauk SSSR **224**, No 6, 1256-1259 (1975); Englist transl.: Soviet math. Dokl. **16**, No 5. 1384-1388 (1976)
[31] Saadoune, M.: Une nouvelle extension en dimension infinie du Théorème de Komlós. Application: Compacité faible dans L^1_X, convergence en mesure, Preprint, Université Ibnou Zohr, Agadir, Morocco 1995
[32] Saadoune, M.: Compacité, Convergences et Approximations, Thèse de Doctorat d'Etat, Faculté des Sciences de Rabat, Juin 1996
[33] Slaby, M.: Strong convergence of vector-valued pramarts and subpramarts, Probability and Math. Stat. **5**, 187-196 (1985)
[34] Talagrand, M.: Weak Cauchy sequences in $L^1(E)$. Amer. J. Math. **106**, 703-724 (1984)
[35] Ülger, A.: Weak compactness in $L^1(\mu, X)$. Proc. Amer. Math. Soc. **103**, 143-149 (1991)
[36] Valadier, M.: Convergence en mesure et optimisation, Travaux du Séminaire d'Analyse convexe, Univ. Montpellier II (1976), exp 14.

Adv. Math. Econ. 1, 39–67 (1999)

Product differentiation and market power*

Egbert Dierker and Hildegard Dierker

Institut für Wirtschaftswissenschaften, Universität Wien, Hohenstaufengasse 9, A-1010 Vienna, Austria

Received: April 7, 1998
JEL classification: D43, L13

Abstract. Assuming symmetry across firms and constant unit costs Perloff and Salop (1985) show: If product differentiation increases, prices rise in a symmetric equilibrium. This raises the question of whether, in general, more product differentiation leads to higher market prices. Giving up the symmetry and the constant unit costs assumptions we present examples in which at least one firm lowers its equilibrium price when product differentiation increases. We formulate a model of product differentiation and state and discuss, within the theory of supermodular games, conditions ensuring that all firms raise their prices in a Nash equilibrium if product differentiation increases.

1. Introduction

Consider a duopoly in which Bertrand competition prevails, i.e. each firm sets the price for its product. It is well known that these firms have no market power at all, if the product is homogeneous and firms have identical unit costs c. In this case, there exists a unique Nash equilibrium and the market price equals c in equilibrium due to the strong incentive to undercut each other's price as long as prices exceed unit costs.

If product differentiation is introduced and each firm produces a different brand, then competition is mitigated and one might conjecture that a higher degree of product differentiation entails a lower degree of competitiveness on the market. That is to say, if product differentiation increases equilibrium prices go up. Indeed, Perloff and Salop (1985) introduce a model in which the degree of product differentiation can be parametrically varied and the above conjecture holds true in the following specific sense: If symmetry with respect to consumers' demand as well as with respect to unit costs holds and unit costs are constant, then a higher degree of product differentiation entails higher markups in a symmetric equilibrium. This raises the following

* We are grateful to B. Grodal for many valuable discussions and would like to thank R. Amir, G. Götz, and M. Nermuth for helpful comments.

question: Does the conclusion of Perloff and Salop hold more generally or does it crucially depend on the specific setting, in particular, the symmetry of the model (and of the equilibrium under consideration) and the assumption of constant unit costs?

The economic intuition underlying the connection between product differentiation, market power, and returns to scale has been expressed by Anderson et al. (1992), p.1 f., in their comprehensive account of the theory of oligopolistic markets for differentiated products as follows: "Once it is recognized that consumers have idiosyncratic preferences, it follows that they are prepared to pay more for the variants that are better suited to their own tastes. It is these premia that are the source of market power for firms. Despite the wide diversity of tastes, the market is unlikely to support a large number of products because of increasing returns to scale in research and development, production, marketing and distribution. Such increasing returns are necessary for firms to exist." Moreover, they write [see p. 198]: "Considering *economies of scale in conjunction with taste diversity then implies that firms have some degree of market power* in the sense that any individual firm would not lose all its customers by changing its price slightly$\cdots$, since some consumers will be willing to pay a premium for variants that better match their tastes." These statements certainly suggest to relax the assumption of constant units costs. Furthermore, symmetry is nowhere mentioned in these intuitive arguments.

In this paper we give up the symmetry assumption with respect to consumers' demand and with respect to costs. Furthermore, we consider the case of strictly decreasing marginal costs [1]. More precisely, we shall address the following questions:

i) How robust is the monotonicity result by Perloff and Salop (1985) and what does a more general theorem look like?
ii) What is the economic significance of the symmetry assumption underlying this result?
iii) How does the degree of product differentiation affect the existence of Nash equilibria if costs are strictly concave?

A natural framework for the comparison of Nash equilibria, which lies at the heart of the result by Perloff and Salop, is provided by the theory of supermodular games that has been developed in recent years. This theory is based on seminal results by Tarski (1955) and by Topkis (1978, 1979), who also pointed out explicitly the usefulness of the concept of supermodularity

[1] A major part of the literature allowing for increasing returns to scale does so by adding a fixed cost component to a linear variable cost function. This is done in order to study questions concerning the number of firms active in equilibrium and the variety of products offered. It turns out, however, that some results (e.g. the dependence of the equilibrium prices on the number of firms on the market) change when strictly concave cost functions are introduced.

for the investigation of various questions arising in economic theory and non-cooperative game theory. Their work constitutes an important cornerstone of modern mathematical economics and related areas. More recent, valuable contributions have been made by Vives (1990), Milgrom and Roberts (1990), and Milgrom and Shannon (1994). Basic definitions and results that are relevant for the present analysis are summarized in Section 2.1. Moreover, in Section 2.2 we present a model of horizontal product differentiation in the tradition of discrete choice theory. This model is based on the intuition that the competition for customers through undercutting is the central feature of oligopoly games with price setting firms. Since the degree of product heterogeneity can easily be varied, the model enables us to compare equilibrium prices for different degrees of product differentiation.

In Section 3 some examples of Bertrand competition are presented in which the equilibrium price of at least one firm behaves counterintuitively. That is to say, in these examples some firm lowers its price in equilibrium if product differentiation increases. Furthermore, we also give an example in which *all firms exhibit counterintuitive behavior.* This particular example exploits the presence of economies of scale. An extensive discussion of these examples provides an explanation of the economic nature of a formal condition that arises naturally in the context of supermodular games and is made in order to rule out counterintuitive behavior.

In Section 4.1 we shall present a local result that applies to linear as well as to other cost functions. Assume that the game under consideration is (locally) supermodular and that the economic reasons for counterintuitive behavior discussed in Section 3 are absent. Then it is still not true that equilibrium prices must rise if product differentiation increases. It turns out that a dominant diagonal condition is necessary and sufficient to obtain this conclusion. This fact is derived using arguments familiar from the theory of Leontief models as developed, e.g., in Nikaido (1968).

It is well known that dominant diagonal conditions can be employed to show uniqueness of equilibrium. Thus, their highly restrictive nature is apparent, but the economic meaning of a dominant diagonal assumption needs clarification. For that reason, we shall place ourselves in a more specific setting in Section 4.2. In particular, we assume constant marginal costs and formulate a dominant diagonal condition in terms of the monotonicity of a certain demand elasticity. As a result, our main assumptions used to show the existence and uniqueness of Bertrand-Nash equilibria (ruling out concave cost functions) are all expressed in terms of demand elasticities. Furthermore, under the assumptions made equilibrium prices rise if products become more heterogeneous. However, equilibrium may fail to exist if marginal costs are falling. In Sections 5 and 6 we look at the question of whether, in case of concave costs, a sufficiently high degree of product differentiation can restore the supermodularity or the quasiconcavity, respectively, of a firm's profit function. We show that quasiconcavity can be achieved if the cost function is not too concave. More precisely, we present an inequality relating the elasticity of the marginal cost function to the required degree of product heterogeneity. We argue that such a clear-cut relation cannot be established for supermodularity.

2. The mathematical framework

2.1 Topkis' monotonicity theorem and smooth supermodular games

The vast majority of the equilibrium existence proofs in economics is based on Brouwer's fixed point theorem or one of its generalizations, e.g., Kakutani's theorem. However, to incorporate issues of comparative statics as discussed in the introduction, it is often useful to shift the focus from topology to partial orders and lattices [2]. A nonempty set S together with a partial order $\leq$ is a lattice if each pair x and y of points in S has a supremum denoted $x \vee y$ and an infimum denoted $x \wedge y$ in S. A lattice $(S, \leq)$ is complete, if each nonempty subset $S' \subset S$ has a supremum $\bigvee S'$ and an infimum $\bigwedge S'$ in S. A function f from a partially ordered set S into a partially ordered set T is called increasing or isotone if $x \leq y$ implies $f(x) \leq f(y)$. The existence of equilibria in lattice theoretical models follows from Tarski (1955).

Tarski's Fixed Point Theorem: Let $(S, \leq)$ be a complete lattice, $f : S \to S$ be increasing, and F be the set of fixed points of f. Then $F \neq \emptyset$ and $(F, \leq)$ forms a complete lattice. Moreover, $\bigvee F = \sup\{x \in S \mid f(x) \geq x\}$ is the largest and $\bigwedge F = \inf\{x \in S \mid f(x) \leq x\}$ is the smallest fixed point of f.

The importance of lattice-theoretical fixed point theorems for economics, game theory and related fields such as operations research has been pointed out by Topkis in several publications that had a great impact on economic theory. In his paper on equilibrium points in submodular games Topkis (1979) mentions explicitly "the pricing problem of competitors producing substitute goods" as an application. Topkis (1978) is of particular relevance, since this "paper gives general conditions under which a collection of optimization problems, with the objective function and the constraint set depending on a parameter, has optimal solutions that are an isotone function of the parameter." The close connection to the problem dealt with here becomes clear if one considers the best response of a firm to the strategies of its competitors in a family of oligopoly problems parameterized by the degree of product differentiation.

A real-valued function f defined on a lattice $(S, \leq)$ is *supermodular* if $f(x) + f(y) \leq f(x \vee y) + f(x \wedge y)$ for all $x, y \in S$. A real-valued function

[2] A partial order is a binary relation that is reflexive, transitive and antisymmetric.

g defined on the product $S \times T$ of two partially ordered sets has *increasing differences* if $g(x,t) - g(x,t')$ is isotone in x for each $t' < t$. Since $g(x,t) - g(x,t') \leq g(x',t) - g(x',t')$ iff $g(x,t) - g(x',t) \leq g(x,t') - g(x',t')$, there is no distinction between the two factors S and T in the definition.

In our oligopoly problem as well as in many other applications to economics the underlying space is a finite product of intervals in $\mathbb{R}$ and the function f under consideration is C^2.[3] It is shown in Topkis (1978), section 3, that such a real-valued function f is supermodular if and only if $\partial_j \partial_i f \geq 0$ for $j \neq i$ [4].

Topkis (1978), section 6, studies isotone optimal solutions of a family of optimization problems parameterized by a partially ordered set T. Since he presents three theorems phrased in a terminology that has not been introduced here, it is convenient and sufficient for our purposes to take the following version of Topkis' results from Milgrom and Roberts (1990).

Topkis' Monotonicity Theorem: Let S be a lattice and T a partially ordered set. Suppose $f(s,t) : S \times T \to \mathbb{R}$ is supermodular in s for given t and has increasing differences in s and t. Suppose that $t \geq t'$ and that $s \in M = \operatorname{argmax} f(s,t)$ and $s' \in M' = \operatorname{argmax} f(s,t')$. Then $s \wedge s' \in M'$ and $s \vee s' \in M$. In particular (when $t = t'$), the set of maximizers of f is a sublattice.

We shall consider smooth supermodular games of the following type. Every player $i = 1, \cdots, n$ has a strategy set S_i that is a compact interval in $\mathbb{R}$ [5]. Let $S = \prod_{i=1}^{n} S_i$ be the (complete) lattice of strategy profiles. The family of games is parameterized by an interval $T \subset \mathbb{R}$. The payoff $f_i : S \times T \to \mathbb{R}$ of every player i is C^2 and satisfies the supermodularity condition $\partial_j \partial_i f_i(s;t) \geq 0$ where the mixed partial derivatives are taken with respect to strategies $s_i, s_j, j \neq i$. Furthermore, f_i has increasing differences in i's own strategy s_i and the parameter $t \in T$. That is to say, we require $\partial_t \partial_i f_i(s;t) \geq 0$. Milgrom and Roberts (1990), [cf. Theorem 6 and its corollary] have pointed out that Tarski's fixed point theorem and Topkis' monotonicity theorem entail the following result.

Comparative statics of Nash equilibria in smooth supermodular games: Under the conditions listed above each game has a smallest and a largest Nash equilibrium in pure strategies. Each of them is a nondecreasing function of t.

We say that there is strategic complementarity between firms i and j, if firm i's best response is increasing in firm j's price and vice versa. The link between supermodularity and strategic complementarity is provided by Topkis' monotonicity theorem. Strategic complementarity and supermodularity both

[3] As usual f is supposed to be C^2 on an open neighborhood if the intervals are compact.

[4] For two vectors $x = (x_1, \cdots, x_n)$ and $y = (y_1, \cdots, y_n)$, the relation $x \leq y$ means $x_j \leq y_j$ for $j = 1, \cdots, n$.

[5] Except for notational adjustments no changes are required if S_i is taken to be a compact interval in $\mathbb{R}^{n_i}$.

do not depend on the choice of coordinates in which strategies are described. In contrast to strategic complementarity, supermodularity is not invariant, though, with respect to coordinate transformations of the payoffs. However, at points satisfying the first order condition for an optimal response, the condition $\partial_j \partial_i \varphi(f_i(s;t)) \geq 0$ holds for every coordinate transformation φ with $\varphi' > 0$ if and only if $\partial_j \partial_i f_i(s;t) \geq 0$ [6].

It should be apparent from Topkis' monotonicity theorem that the condition of increasing differences in the strategies s_i and the parameter t plays a crucial role when Nash equilibria corresponding to different degrees of product differentiation are compared. Section 3 sheds light on the economic significance of this requirement by means of examples exhibiting counterintuitive behavior. Furthermore, we shall give a clear meaning to the parameter t measuring product differentiation in subsection 2.2.

2.2 A model of horizontal product differentiation

To address the questions listed in the introduction in a systematic fashion we have to use a theoretical framework in which the notion of the degree of product differentiation can be expressed. It appears natural to us to use a discrete choice model or, more precisely, an additive random utility model [cf. Anderson, de Palma and Thisse (1992), pp. 86-90]. This approach has the advantage that it does neither rely on the existence of a representative consumer nor on the specification of particular utility functions.

In our setting there are n brands, indexed $j = 1, \cdots, n$, of a perfectly divisible product and another divisible (composite) commodity indexed 0, which is needed to close the model. Brand j is produced by firm j out of good 0.

Good 0 will serve as numéraire. The objective of each firm is to maximize profits in terms of good 0. That is to say, we implicitly assume that all shareholders of all firms are only interested in the numéraire. If such a restrictive assumption is not made, one has to deal with the question of why maximization of profits in terms of good 0 reflects the interests of the firm's shareholders correctly. As a convenient by-product, we obtain the following consequence. Since shareholders neither consume nor sell any of the brands, firms face demand functions which are independent of the size and distribution of profits.

The rate of substitution between two brands i and j is assumed to be constant for any individual consumer and this constant is distributed across the population according to some probability law. If the distribution is highly concentrated near a certain number (which we can think of as normalized to 1), then the brands i and j are nearly perfect substitutes for the vast majority of customers. If the standard deviation is enlarged, the concentration will

---

[6] For an ordinal definition of supermodularity, see Milgrom and Shannon (1994).

be reduced and the diversity of tastes increases, i.e. products become more differentiated. Thus, the notion of product differentiation can be associated with the concept of a standard deviation of a distribution.

Let the distribution of the rates of substitution be absolutely continuous with respect to Lebesgue measure. Then only a null set of consumers buys more than one brand at a given price system $(1, p_1, \cdots, p_n)$. To be specific, let consumer a's utility be given by

$$u^a(x_0, x_1, \cdots, x_n) = v^a(x_0, \sum_{j=1}^{n} \delta_j^a x_j) \tag{1}$$

where $\delta_j^a > 0$. Given prices $(1, p_1, \cdots, p_n)$, consumer a chooses brand i if $\delta_i^a / \delta_j^a > p_i / p_j$ for all $j \neq i$, because his rate of substitution between brand i and any brand $j \neq i$ then exceeds the price ratio p_i/p_j. In this case the amounts of brand i and of the numéraire demanded at the price system $(1, p_1, \cdots, p_n)$ by consumer a can be derived from the conditional utility function $u_i^a(x_0, x_i) = v^a(x_0, \delta_i^a x_i)$ together with a's budget constraint. For convenience, the demand obtained if a consumer (or a group of consumers) is restricted to buy only a certain brand, say i, in addition to the numéraire is referred to as *conditional demand* for i.

Alternatively, consumer a's taste can be described by a system of functions $u_j^a(x_0, x_j) = v^a(x_0, \delta_j^a x_j)$, $j = 1, \cdots, n$, measuring a's utility in case he consumes only brand j in addition to the numéraire. It is often convenient to use logarithmic coordinates and to write $u_j^a(x_0, x_j) = g^a(\log(x_0), \log(x_j) + \epsilon_j^a)$, where $\epsilon_j^a = \log(\delta_j^a)$. In Section 3 we shall analyze some particular examples in which u_j^a is Cobb-Douglas. In this case, we obtain $u_j^a(x_0, x_j) = \log(x_0) + \log(x_j) + \epsilon_j^a$ and the random vector $\epsilon = (\epsilon_1, \cdots, \epsilon_n)$ enters additively (additive random utility model). Of course, firms don't have to know the underlying stochastic model, but they are supposed to correctly anticipate their demand.

It is worth noticing that CES demand functions, which are often used in models of product differentiation, can be obtained as a particular case if the distribution of ϵ is specified as follows [cf. Anderson et al. (1992), pp. 85-90]. Let $\epsilon_1, \cdots, \epsilon_n$ be i.i.d. random variables with distribution function given by the double exponential expression $F(x) = \exp\{-\exp(-(\gamma + x/\mu))\}$, where $\gamma \approx 0.577$ is Euler's constant and μ is, up to a positive factor, the standard deviation [7]. A calculation yields that brand i is chosen with probability $p_i^{-1/\mu} / (\sum_{j=1}^{n} p_j^{-1/\mu})$. Clearly, demand is split across brands as if there were a representative consumer with a CES utility for the brands. Given the Cobb-Douglas utility specification according to which the expenditure share for the

[7] This distribution plays an important role in applied models of discrete choice (so-called logit models). It has the following two rather convenient properties. Let ϵ_1 and ϵ_2 be independently distributed according to the double exponential. Then $\max\{\epsilon_1, \epsilon_2\}$ has a distribution of the same type and $\epsilon_1 + \epsilon_2$ has a logistic distribution.

brands is fixed[8] in the additive random utility model described above, the demand for brand i becomes

$$f_i(p_1, \cdots, p_i, \cdots, p_n) = \frac{1}{p_i} \frac{p_i^{-1/\mu}}{\sum_{j=1}^{n} (p_j^{-1/\mu})} . \tag{2}$$

The limit case where the CES utility function becomes linear and brands become perfect substitutes corresponds to $\mu \to 0$, i.e. to a vanishing standard deviation. Conversely, if $\mu \to \infty$, the CES utility becomes Cobb-Douglas and firm i's demand is unaffected by the prices charged by other firms.

It should be obvious that the degree of product differentiation can naturally be expressed in any stochastic discrete choice model in the same way as described in the context of the CES example. The standard deviation of the distribution of any random vector ϵ can be arbitrarily scaled up and down by choice of the parameter $\mu > 0$, if one uses the mapping $\epsilon \mapsto \mu\epsilon$. If μ goes to 0 and ϵ is centered at 0, the distribution of $\mu\epsilon$ approaches (weakly) the Dirac measure concentrated at 0. Then there is no product differentiation at all and we are in the (degenerate) case of pure Bertrand competition for a homogeneous product. Assume now that ϵ has a continuous density. Then, for μ tending to ∞, infinitesimally small price changes Δp_j of firm j have negligible effects on the market share of any firm at any price system, since the density is bounded by an arbitrarily small number for μ sufficiently large. Therefore, cross-price effects $\partial_j f_i, j \neq i$, vanish for $i, j \geq 1$ as in case of Cobb-Douglas demand. *Perfect product differentiation* prevails in the sense that (bounded) price variations of firm j cease to have an impact on the profits of any competitor $i \neq j$. Observe that this fact is independent of the shape of the underlying probability density of ϵ provided that this density is bounded. In particular, stochastic independence is irrelevant here. For a very large degree of product differentiation one is bound to be approximately in the Cobb-Douglas case [cf. Grandmont (1992)].

In the sequel we assume implicitly that product differentiation is described in terms of a discrete choice model as presented above, a distribution with a higher standard deviation corresponding to a higher degree of product differentiation. Although the formal analysis will be independent of the details of this description, its qualitative nature is of great importance for the understanding of the relationship between product differentiation and market power.

3. Counterintuitive behavior of firms

In all three examples given below there are two price setting firms producing brand 1 and 2, respectively, of a differentiated product by means of good

[8] For simplicity, we assume that the aggregate expenditure on all brands is normalized to 1.

0, which is used to measure profits. Consumers' aggregate demand for the brands is given by a CES demand function. In Section 2.2 a stochastic model has been presented that gives rise to an aggregate consumer with CES demand. The degree of product differentiation has been expressed in terms of the standard deviation of the underlying distribution. Clearly, product differentiation can as well be measured directly in terms of the parameter in the CES utility function that indicates the elasticity of substitution, if one conceives of the consumption sector as consisting of a continuum of identical CES consumers.

Example 1 (demand asymmetry): Consumers' valuation of the brands is asymmetric, because brand 2 is given three times as much weight as brand 1. More precisely, consider the utility function

$$u(x_0, x_1, x_2) = x_0 \cdot (0.25x_1^{\beta} + 0.75x_2^{\beta})^{\frac{1}{\beta}}, \tag{3}$$

where $0 < \beta \leq 1$. Total wealth of all customers on the market for the differentiated product equals two and expenditures are split evenly between the numéraire and the brands. Putting $r = \beta/(1-\beta)$ the aggregate demand $f = (f_1, f_2)$ for both brands obtains as

$$\begin{aligned} f_1(p_1, p_2) &= \frac{0.25^{r+1} p_1^{-r}}{p_1(0.25^{r+1} p_1^{-r} + 0.75^{r+1} p_2^{-r})} \\ f_2(p_1, p_2) &= \frac{0.75^{r+1} p_2^{-r}}{p_2(0.25^{r+1} p_1^{-r} + 0.75^{r+1} p_2^{-r})} \, . \end{aligned} \tag{4}$$

We shall refer to firm 1 as the weak and to firm 2 as the strong firm.

If $\beta = 1$, i.e. if $r = \infty$, preferences are linear and the elasticity of substitution between brands is infinite. This corresponds to the case of pure Bertrand competition without any product differentiation at all. On the other hand, if β and hence r decrease, the elasticity of substitution does too, that is to say, product differentiation increases. In the limit, if r tends to 0, utility becomes Cobb-Douglas and product differentiation becomes perfect since a change of the price of firm i does not at all affect the expenditures on brand j for $j \neq i$. Thus, $t = 1/r$ is a measure of product differentiation[9].

In this example, firms have identical constant unit costs $c = 1$ so that the profit of firm i is given by $\Pi_i(p_1, p_2) = (p_i - 1) f_i(p_1, p_2), i = 1, 2$. A numerical computation yields equilibrium prices $p_1 \approx 1.0693$, $p_2 \approx 2.7798$ for $t = 1/15$. For $t = 1/14$ equilibrium prices become $p_1 \approx 1.0744$, $p_2 \approx 2.7791 < 2.7798$. Thus, the strong firm exhibits counterintuitive behavior. Furthermore, it turns out that not only firm 2's profits, but also aggregate

[9] Observe that t coincides with μ that has been used in Section 2.2 for the same purpose.

profits, decline, if the degree of product differentiation increases from 1/15 to 1/14.

Moreover, if t rises further to 1/13, the behavior of the strong firm 2 is reversed, since its price starts to rise slightly to $p_2 \approx 2.7794$. On the other hand, computing the equilibrium values of p_1 we found no conspicuous results. Thus we are led to ask:

a) Why is it the strong rather than the weak firm that exhibits counterintuitive behavior?
b) Why is the behavior of the strong firm reversed if t gets larger than some critical value (around 1/14 in this example)?

To give an intuitive answer to these questions it is worthwhile to look at the behavior of equilibrium prices for all positive t. If the degree t of product differentiation is very close to zero, then the corresponding CES indifference curves are rather flat with normal vector close to $(1, 3)$ (apart from a very small area close to the boundary of $\mathbb{R}^2_+$ where the indifference curves bend towards the horizontal and vertical coordinate axis, respectively). As a consequence, the weak firm's price p_1 is nearly equal to its unit costs $c = 1$ and the strong firm charges approximately three times as much. The weak firm has a negligible market share and the strong firm is nearly a monopolist restricted, though, to some extent by the presence of the weak firm. If the degree t of product differentiation is still low, but begins to rise, the weak firm starts getting a larger market share and the strong firm becomes less dominant. As a result, the weak firm can raise its price. The strong firm, however, suffers from the increase in competition. Losing part of its monopoly power, the strong firm has to lower its price.

On the other hand, if t gets very large, the Cobb-Douglas case with expenditure shares 1/4 and 3/4, respectively, is approximated. Both firms sell small, but roughly equal amounts and the price ratio p_1/p_2 is slightly above 1/3. In particular, a larger t yields an increase in market power for both firms, since they split the market in nearly fixed proportions.

Summarizing, the price of the strong firm decreases with t for small values of t, whereas it rises with t for large ones. The economic explanation is the following. There are two different types of market power in the example. For $t \approx 0$, demand asymmetry puts the strong firm into the position of a restricted monopolist. If t increases, consumers become more mobile thereby weakening the position of the strong firm. Finally, both firms become more powerful if t gets sufficiently large. The latter case corresponds to the intuition expressed by Anderson et al. (1992) as cited in the introduction. Their discussion focuses on the market power all firms together exert over the consumers, but it abstracts from the question of how the strategic position of an individual firm in relation to its competitors is affected by the degree of product differentiation.

Example 2 (small cost asymmetry): The above discussion suggests that counterintuitive behavior for sufficiently low degrees t of product differentiation can occur independently of the size and of the source of the underlying asymmetry. That is to say, the monotonicity result by Perloff and Salop is likely to show little robustness. To illustrate this point, we are now going to give an example in which unit costs rather than demand functions are asymmetric and the strong firm is given a cost advantage of only 1 percent[10].

In Example 2 the utility function is symmetric with respect to brands and given by

$$u(x_0, x_1, x_2) = x_0 \cdot (0.5x_1^\beta + 0.5x_2^\beta)^{\frac{1}{\beta}}. \tag{5}$$

As before total wealth of all customers equals 2. Putting again $r = \beta/(1-\beta)$ the aggregate demand f_i for brand $i = 1, 2$ becomes

$$f_i(p_1, p_2) = \frac{p_i^{-r}}{p_i(p_1^{-r} + p_2^{-r})} \ . \tag{6}$$

Unit costs are constant and equal $c_1 = 1$ for firm 1 and $c_2 = 0.99$ for firm 2.

A numerical calculation yields that the strong firm charges $p_2 \approx 0.999101$ in equilibrium if $t = 1/2000$. This price declines until t reaches approximately $1/1200$ where $p_2 \approx 0.999006$ and then starts to increase again. Aggregate profits in equilibrium continue to fall until t has reached about $1/700$ and then start growing. The interpretation of this phenomenon is the same as in the previous example: Close to the case of pure Bertrand competition the monopoly power of the strong firm decreases if product differentiation increases whereas the weak firm becomes more active.

We are now going to investigate this phenomenon in more detail by analyzing the behavior of marginal profits $\partial_i \Pi_i = (p_i - c_i)\partial_i f_i + f_i$ for $i = 1, 2$. Let equilibrium prices $p = (p_1, p_2)$ be fixed and observe that the cost advantage of the strong firm 2 induces it to offer its product at a price p_2 which is a bit below p_1. Now consider an infinitesimally small increase Δp_2 of the price of the strong firm. Clearly, Δp_2 entails a decline of size $|\Delta f_2|$ in firm 2's demand. If t is nearly zero, $|\Delta f_2|$ is very small, since consumers consider both products as roughly identical and very few of firm 2's customers at prices p

[10] We would like to remark that example 1 can be transformed into an example with a symmetric CES function and asymmetric costs. Put $\bar{x}_1 = (1/2)^{1/\beta} x_1$ and $\bar{x}_2 = (3/2)^{1/\beta} x_2$. The utility function then becomes

$$u(\bar{x}_1, \bar{x}_2) = x_0 \cdot (0.5\bar{x}_1^\beta + 0.5\bar{x}_2^\beta)^{1/\beta} .$$

To compute an equilibrium in example 1 in terms of the new economy with symmetric demand, observe that unit costs are transformed into $\bar{c}_1 = (1/2)^{-1/\beta}$ and $\bar{c}_2 = (3/2)^{-1/\beta}$. The equilibrium prices (p_1, p_2) in the original economy can be obtained from their counterpart by use of the following transformation: $p_1 = (1/2)^{1/\beta}\bar{p}_1$ and $p_2 = (3/2)^{1/\beta}\bar{p}_2$.

are willing to pay the additional amount of $p_1 - (p_2 + \Delta p_2) > 0$ to get brand 1 rather than brand 2. However, if the degree t of product differentiation gets larger, more customers are willing to forego this incremental amount to obtain their preferred brand 1. That is to say, $|\Delta f_2|$ grows if t does due to the fact that the price advantage of the strong firm loses part of its significance if t rises. Since $\Delta f_2 < 0$, we conclude that $\partial_t \partial_2 f_2(p)$ is negative for sufficiently small t.

There is a second effect which deserves to be discussed. *The loss of market power of the strong firm is accompanied by a reduction of its market share*, since the weak firm gains more customers. Therefore, firm 2 is characterized by the following two properties. For low degrees of product differentiation the inequalities $\partial_t \partial_2 f_2(p) < 0$ and $\partial_t f_2(p) < 0$ hold both at a Nash equilibrium p. Since unit costs are assumed to be constant in the present setting, these conditions imply that marginal profits $\partial_2 \Pi_2(p) = (p_2 - c_2)\partial_2 f_2(p) + f_2(p)$ are locally decreasing with respect to t.

The situation of the weak firm is harder to understand, since the changes $\partial_t \partial_1 f_1$ and $\partial_t f_1$ go into opposite directions for small values of t. Clearly, the positive sign of $\partial_t f_1$ reflects firm 1's gain in customers. We are now going to argue that $\partial_t \partial_i f_i$ tends to have the *same sign for both firms* if t is small. For small t equilibrium prices p_1 and p_2 are both nearly equal to the maximum of the unit costs. Therefore, if both firms raise their prices infinitesimally by the same amount Δp, the relative price of the two brands stays approximately unaltered and customers don't switch firms. Thus, if firm i raises its price unilaterally by Δp and thereby drives some customers away to the other firm j, then these customers will come back to i, if j also raises its price by the same amount Δp. As a consequence, if i's loss of customers caused by a unilateral price increase is higher at $t + \Delta t$ than at t, then the same holds true for j.

Observe, however, that the influence of $\partial_t \partial_i f_i < 0$ on firm i's marginal profits $\partial_i \Pi_i$ is weighted with i's markup $p_i - c_i$, which is much smaller for firm 1 than for firm 2. Actually, it turns out that the other effect $\partial_t f_1 > 0$ reflecting 1's increase in market share is dominant so that $\partial_t \partial_1 \Pi_1$ is positive in a Nash equilibrium.

This fact can be explained by considering the elasticity of the demand f_1. First observe that changes in profits Π_i and elasticities ε_i of demand for brand i are related as follows. Since $\log \Pi_i(p;t) = \log(p_i - c_i) + \log f_i(p;t)$, we have $\partial_t \partial_i \log \Pi_i(p;t) = \partial_t \partial_i \log f_i(p;t)$. Therefore, $\partial_t \partial_i \log \Pi_i(p;t)$ and $\partial_t \varepsilon_i(p;t)$ have the same sign. In Example 2 the elasticity of the demand f_1 of the weak firm obtains at t as

$$\varepsilon_1(p;t) = -1 - \frac{p_1^{1/t}}{(p_1^{1/t} + p_2^{1/t})t} = -1 - \frac{1}{(1 + (\frac{p_2}{p_1})^{1/t})t} . \tag{7}$$

Since in equilibrium the price p_1 of the weak firm in the present example is above that of the strong firm, $(p_2/p_1)^{1/t}$ and hence $(1 + (p_2/p_1)^{1/t})t$ increase in t. Therefore, $\varepsilon_1(p;\cdot)$ and hence $\partial_1 \log \Pi_1(p;\cdot)$ is increasing with respect to t at equilibrium prices.

It remains to discuss the case of the strong firm for large values of t. If t gets larger, the strong firm continues to lose some of its demand, i.e. $\partial_t f_2(p;t) < 0$, but, since it is no longer a restricted monopolist, $\partial_t \partial_2 f_2(p;t) \geq 0$. Considering ε_2 we observe that $(p_1/p_2)^{1/t} t$ is increasing in t for $t > \log(p_1/p_2)$. Hence marginal profits $\partial_2 \Pi_2$ are increasing for t large enough provided the price ratio p_1/p_2 stays bounded at Nash equilibria. Since p_1/p_2 stays around 1.01 in example 2 (and around 3 in example 1), the profit peak of the strong firm shifts also to the right at a Nash equilibrium if t is sufficiently high.

To shed light on the monotonicity result by Perloff and Salop (1985), abstract for the moment from the differences in unit costs and consider a symmetric price system p. The market share of firm i is given by $s_i(p_1, p_2; t) = p_i^{-1/t}/(p_1^{-1/t} + p_2^{-1/t})$. Observe that the elasticity ε_i of i's demand equals the elasticity of i's market share s_i minus 1. The symmetric case is special in the sense that market shares at p are independent of t, because firms split the market evenly. Therefore, in the symmetric case, $\partial_t \varepsilon_i(p;t) > 0$ iff $\partial_t \partial_i s_i(p;t) > 0$. Since the loss of customers $\partial_i s_i(p;t)$ due to an increase in p_i is inversely related to t, $\varepsilon_i(p;\cdot)$ and hence marginal profits $\partial_i \Pi_i(p;\cdot)$ must be increasing for every firm. In the particular setting of this example the elasticity of s_i is $1/(2t)$ whenever $p_1 = p_2$ and we obtain:

Remark 1. Consider the demand functions underlying example 2, equal unit costs, and symmetric equilibrium prices $p_1 = p_2$. Then $\varepsilon_i(p;t) = -1 - \frac{1}{2t}$ is increasing in t. Hence marginal profits $\partial_i \Pi_i(p;\cdot)$ are increasing in t for $i = 1, 2$.

It is apparent in the setting of Remark 1 that the result is driven by two facts. First, market shares are constant, i.e. $\partial_t f_1 = \partial_t f_2$ at Nash equilibria. Second, product differentiation is the only source of market power, hence $\partial_1 f_1$ and $\partial_2 f_2$ are increasing in t at Nash equilibria.

Example 3 (counterintuitive behavior of all firms): To obtain an example in which both firms exhibit counterintuitive behavior, we enrich the previous framework by another economic phenomenon that is pertinent to models of product differentiation. As pointed out in the introduction, the assumption of increasing returns to scale is needed in order to explain that only a restricted spectrum of brands is offered on the market although consumers have a wide diversity of tastes. In most of the relevant literature, increasing returns enter in the form of fixed costs combined with constant marginal costs. However, fixed costs drop out when the profit function is differentiated once. Hence they only affect the level of equilibrium profits but not equilibrium prices and markups. Therefore, it is desirable to investigate the case of concave costs.

In Example 3 demand is symmetric and the same as in Example 2. Marginal costs are decreasing for both firms and asymmetric. More precisely, the cost functions of firms 1 and 2 are, respectively,

$$C_1(x_1) = 3x_1{}^{\frac{9}{10}} \qquad \text{and} \qquad C_2(x_2) = x_2{}^{\frac{9}{10}} . \tag{8}$$

Note that the profit functions Π_i are not necessarily quasiconcave with respect to p_i. Moreover, firms may make losses in equilibrium and, therefore, prefer to stay out. Thus we proceed as follows. The first order conditions for a Nash equilibrium are solved for different values of t and the corresponding profits are computed. One observes that a sufficiently large degree t of product differentiation does not only entail positive profits at the solution, but also the quasiconcavity of both profit functions given the opponent's equilibrium price. Thus, Nash equilibria are obtained for sufficiently large t in which no firm regrets to have entered the market.

For $t = 1/10$ the weak firm makes losses. However, if we take $t = 1/9$, both firms have positive profits and we obtain the equilibrium prices $p_1 \approx 4.784$ and $p_2 \approx 3.427$. Now, if t increases further to $1/8$, equilibrium prices become $p_1 \approx 4.778$ and $p_2 \approx 3.348$. Thus both firms exhibit counterintuitive behavior.

The economic explanation is as follows. As in the first two examples the strong firm 2 loses part of its power due to its cost advantage, if t goes up from $1/9$ to $1/8$ and the weak firm 1 gains strength. However, observe that the weak firm is handicapped twice in this setting. First, for any given output x, costs of the weak firm are three times as high as those of the strong firm. Therefore, the weak firm has to charge more than the strong one and hence its sales are lower. This fact, however, entails another disadvantage of firm 1, since marginal and unit costs increase for diminishing sales. As a consequence, if product differentiation increases and the weak firm sells more, it is induced to lower its price.

The impact of the concavity of the cost function on the behavior of firm 2 can be seen more clearly as follows. Marginal profits at $p = (p_1, p_2)$ are $\partial_i \Pi_i(p) = f_i(p) + p_i \partial_i f_i(p) - C_i'(f_i(p)) \cdot \partial_i f_i(p)$. Hence,

$$\partial_t \partial_i \Pi_i(p) = (p_i - C_i'(f_i(p)) \cdot \partial_t \partial_i f_i(p) + (1 - C_i''(f_i(p)) \partial_i f_i(p)) \cdot \partial_t f_i(p) . \tag{9}$$

We have to compare two effects: The change in market power due to a reduction of consumers' loyalty and the change in market share, each caused by a slight increase in t. The first effect is reflected by the fact that, for the strong firm, $\partial_t \partial_2 f_2(p) \approx -0.43 < 0$ at the equilibrium corresponding to $t = 1/9$. As argued above, the corresponding term $\partial_t \partial_1 f_1(p)$ of the weak firm tends to have the same sign. Indeed, we have $\partial_t \partial_1 f_1(p) \approx -0.33 < 0$. According to the above formula, these negative terms are weighted with the markups $p_1 - C_1' \approx 0.5$ and $p_2 - C_2' \approx 2.4$, respectively, and must then be combined with the second effect.

The strong firm loses some of its market share and we have $\partial_t f_2(p) \approx -0.36$. The weak firm gains customers. Thus, $\partial_t f_1(p) \approx 0.25$ is positive. The remarkable aspect in Example 3 is the following. The weight attached to the second effect depends on the concavity of the cost function of the firm under consideration together with the reaction $\partial_i f_i(p)$ of the firm's demand to a change in its price. Calculating the numbers $C_i''(f_i(p)) \cdot \partial_i f_i(p)$ we obtain 0.86 for the weak firm 1 and 0.04 for firm 2. Since the scale of operation of the strong firm is sufficiently high, the absolute value of C_2'' is low and the concavity of C_2 has little influence. However, in case of the weak firm, the situation is totally different. The two effects $\partial_t\partial_1 f_1(p) < 0$ and $\partial_t f_1(p) > 0$ have opposite signs and the weight assigned to $\partial_t f_1(p)$, which would be 1 in case of constant marginal costs, is reduced to $1 - 0.86 = 0.14$ due to the concavity of C_1 and the low level of firm 1's demand. As a consequence, the first effect dominates the second and $\partial_t\partial_1 \Pi_1(p)$ becomes negative. Therefore, the weak firm behaves in the same way as a restricted monopolist and lowers its price. [11]

As pointed out above, our examples rely on the interplay of two types of market power: The power of a restricted monopolist and the one derived from product differentiation. Anderson and de Palma (1996) also study the interplay of two types of market power. They consider n firms located equidistantly on a circle. Each firm offers one brand of a divisible product. Consumers are uniformly distributed on the circle and transportation costs are independent of the amount bought. If all brands are identical, each consumer buys from that firm for which the price plus transportation cost is the lowest. As brands become more differentiated, consumers are willing to pay more for their preferred brand. Finally, if the degree of product differentiation is very large, transportation costs can be neglected and competition is fully global. Focusing on symmetric equilibria, Anderson and de Palma (1996) show that equilibrium prices first decrease if product differentiation increases. But, if product differentiation gets strong enough, equilibrium prices rise with the degree t of product differentiation. In their setting the behavior of marginal profits is solely determined by $\partial_t\partial_i f_i$, since $\partial_t f_i$ is identically equal to 0 due to the symmetry assumption (fixed market shares $1/n$). Clearly, if customers start shopping around and globalization of competition sets in, the decisive term $\partial_t\partial_i f_i$ becomes negative and firms lose market power. For sufficiently large t, there is nearly no local market power left and all firms benefit from a further increase of product differentiation.

[11] In the intuitive discussion of examples we abstract from the indirect effect a price change of firm i exerts on firm $j \neq i$. In Section 4 we study Nash equilibria and assume strategic complementarity to control the interaction among firms.

4. When does more product differentiation entail higher market prices?

4.1 A local result for arbitrary returns to scale

It is apparent from the discussion of Example 3 in the previous section that Bertrand-Nash equilibria need not exist if marginal costs are strictly decreasing. This is due to the fact that both types of fixed point theorems (based on Brouwer's and Tarski's approach, respectively) break down in case of sufficiently concave cost functions, since quasiconcavity as well as strategic complementarity then cease to hold.

In this subsection we shall, therefore, disregard the existence issue and assume that the economy under consideration has an equilibrium p^* for a certain degree $\bar{t}$ of product differentiation. We shall investigate the following question. Assume:

a) each individual firm i has an incentive to raise its price if t grows, that is to say, $\partial_t \partial_i \Pi_i(p^*; \bar{t})$ is positive for each i;
b) for given $\bar{t}$ the economy exhibits strategic complementarity at the equilibrium price system p^*.

Under which conditions do equilibrium prices rise if t is slightly increased beyond its original level $\bar{t}$?

The answer to this question sheds light on the role of strategic complementarity in comparative statics. In particular, the answer puts the result on the behavior of the largest and the smallest equilibrium in Milgrom and Roberts [1990], that has been quoted in Section 2.1, into perspective.

To analyze the question stated above it suffices to use the following flexible setting, in which the payoff Π_i can be thought of as an isotone transformation of i's profits. The strategy space of each firm i is a one dimensional interval S_i representing prices. A larger element of S_i stands for a higher price.

Assume now that $s^* \in S = \prod_{i=1}^n S_i$ is an equilibrium for the given degree $\bar{t}$ of product differentiation and let $\partial_t \partial_i \Pi_i(s^*; \bar{t})$ be positive for every firm $i = 1, \cdots, n$. Then each individual firm i has an incentive to raise its price if product differentiation increases, provided i ignores the fact that its competitors change their strategies, too. Clearly, these incentives will be reinforced by the strategic interaction across firms if strategic complementarity prevails, that is to say, if $i's$ optimal response rises if some firm $j \neq i$ increases its price. If strategic complementarity would be violated, then the direct effect induced by $\partial_t \partial_i \Pi_i(s^*; \bar{t}) > 0$ on i could be outweighed by the cross effects associated with strategy increments of i's rivals. Thus, without the assumption of strategic complementarity no clear-cut prediction that does not rely upon comparing counteracting forces seems possible. Strategic complementarity is captured locally by the condition $\partial_j \partial_i \Pi_i \geq 0$ for $j \neq i$ (weak local supermodularity). [12]

However, it is worth emphasizing that the assumption of *strategic complementarity is insufficient* to yield the result that more product heterogeneity implies higher market power although it entails an additional drive to increase prices. Focusing only on the forces just discussed one ignores the fact that a Nash equilibrium is, ultimately, a solution of a system of equations that is not necessarily reached as a limit point of the dynamical system we are sketching now. More precisely, let $p^*(\bar{t})$ be an equilibrium for given $\bar{t}$. If product differentiation increases slightly by $\Delta t > 0$, firms have an immediate incentive to raise their prices if $\partial_t \partial_i \Pi_i(p^*(\bar{t}); \bar{t}) > 0$. This starts a process in which all firms increase their prices at a given stage, because their competitors did so one stage earlier. If strategic complementarity is sufficiently powerful, then there is no reason why the process of successive price increases should converge to the new equilibrium $p^*(\bar{t}+\Delta t)$ that lies near $p^*(\bar{t})$. Indeed, it may be necessary to reverse the dynamics to reach $p^*(\bar{t}+\Delta t)$. Thus, to rule out this type of instability, an additional assumption restricting the degree of strategic complementarity is needed. Otherwise it may very well happen that prices must be lowered to reach the new equilibrium $p^*(\bar{t} + \Delta t)$.

The above intuition will be made precise in the following theorem. To put a bound on the degree of strategic complementarity we shall use the following definition.

Definition. *A matrix $A = (a_{ij})$ of order n has a* dominant diagonal *if there are positive weights $\lambda_1, \cdots, \lambda_n$ such that for all $i = 1, \cdots, n$*

$$\lambda_i |a_{ii}| > \sum_{j \neq i} \lambda_j |a_{ij}| .$$

Obviously, A has a dominant diagonal iff $-A$ does. We shall interpret the entry a_{ij} in this definition as the value of $\partial_j \partial_i \Pi_i(s^*; \bar{t})$. Then $a_{ij} \geq 0$ for $i \neq j$ due to weak local supermodularity. Furthermore, the diagonal terms must satisfy the second order condition $\partial_i \partial_i \Pi_i \leq 0$ at any optimal response. Taking these constraints on the signs of the elements of A into account, the above inequality reads $-\lambda_i a_{ii} > \sum_{i \neq j} \lambda_j a_{ij}$. That is to say, if local supermodularity is assumed, the dominant diagonal condition obtains iff there are positive weights $\lambda_1, \cdots, \lambda_n$ such that the weighted sum of the columns is negative, i.e.

$$\sum_{j=1}^{n} \lambda_j a_{ij} < 0 \qquad \text{for } i = 1, \cdots, n. \tag{10}$$

[12] As noted in Section 2.1, the condition $\partial_j \partial_i \Pi_i \geq 0$ for $j \neq i$ is invariant with respect to increasing transformations of the payoff functions Π_i at i's optimal responses.

In particular, for every firm i, the local degree of concavity of Π_i with respect to i's own strategy s_i as given by $|a_{ii}|$ must be sufficiently large in comparison to the off-diagonal (supermodularity) terms a_{ij}.

The following theorem refers to the setting described above. In particular, s^* denotes an equilibrium for given degree $\bar{t}$ of product differentiation that lies in the interior of the space $S = \prod_{i=1}^n S_i$ of strategy profiles and $\Pi_i(s;t)$ is i's payoff at strategy profile s if the degree of product differentiation equals t.

Theorem 1. *Suppose that the payoffs Π_i are C^2-functions of $(s_1, \cdots, s_n; t)$ and let $s^* \in \operatorname{int} S$ be a Nash equilibrium for $t = \bar{t}$ such that $(\partial_j \partial_i \Pi_i(s^*; \bar{t}))_{i,j=1,\cdots,n}$ has maximal rank. Furthermore, assume*

1) $\partial_j \partial_i \Pi_i(s^*; \bar{t}) \geq 0$ *for $j \neq i$ (weak local supermodularity),*
2) $\partial_t \partial_i \Pi_i(s^*; \bar{t}) > 0$ *for all i (marginal payoffs increase with t).*

Then there exists a neighborhood U of s^* such that, for each t in a neighborhood of $\bar{t}$, the system $\partial_i \Pi_i(\,\cdot\,; t) = 0$, $i = 1, \cdots, n$, of first order conditions for a Nash equilibrium has a unique solution $s^*(t)$ in U. If $\frac{ds^*}{dt}(\bar{t}) \geq 0$, then the matrix $(\partial_j \partial_i \Pi_i(s^*; \bar{t}))$ has a dominant diagonal. Furthermore, if $(\partial_j \partial_i \Pi_i(s^*; \bar{t}))$ has a dominant diagonal, then $\frac{ds^*}{dt}(\bar{t})$ is strictly positive.

The proof is based upon arguments that are well-known from the theory of linear economic models. For a comprehensive, detailed exposition see Nikaido (1968).

Proof. Since s^* is an interior equilibrium, we have $\partial_i \Pi_i(s^*, \bar{t}) = 0$ for $i = 1, \cdots, n$. Define the matrix $A = (a_{ij})$ by $a_{ij} = \partial_j \partial_i \Pi_i(s^*, \bar{t})$. The matrix A is nondegenerate and the implicit function theorem yields the existence of a C^1 function $s^*(\cdot)$ defined on a neighborhood of $\bar{t}$ with values in a neighborhood U of s^* such that $\partial_i \Pi_i(s^*(t); t) = 0$ for $i = 1, \cdots, n$. The function $s^*(\cdot)$ is uniquely determined by $s^*(\bar{t}) = s^*$.

Differentiating the equations $\partial_i \Pi_i(s^*(t), t) = 0$, $i = 1, \cdots, n$, with respect to t we get

$$\sum_{j=1}^{n} \partial_j \partial_i \Pi_i(s^*(\bar{t}), \bar{t}) \frac{ds_j^*(\bar{t})}{dt} + \partial_t \partial_i \Pi_i(s^*(\bar{t}), \bar{t}) = 0 \tag{11}$$

or

$$(-A) \begin{pmatrix} \frac{ds_1^*(\bar{t})}{dt} \\ \vdots \\ \frac{ds_n^*(\bar{t})}{dt} \end{pmatrix} = \begin{pmatrix} \partial_t \partial_1 \Pi_1(s^*, \bar{t}) \\ \vdots \\ \partial_t \partial_n \Pi_n(s^*, \bar{t}) \end{pmatrix}. \tag{12}$$

The vector on the right-hand side is strictly positive by assumption 2). Hence, $-a_{ii} \frac{ds_i^*}{dt}(\bar{t}) - \sum_{j \neq i} a_{ij} \frac{ds_j^*}{dt}(\bar{t}) > 0$ for every i. Suppose $\frac{ds^*}{dt}(\bar{t}) \geq 0$ and put $\lambda_j = \frac{ds_j^*}{dt}(\bar{t}) + \epsilon$, where $\epsilon > 0$. By continuity, ϵ can be chosen small enough to yield $\lambda_i|a_{ii}| > \sum_{j \neq i} \lambda_j |a_{ij}|$ for all $i = 1, \cdots, n$. Therefore, A has a dominant diagonal.

On the other hand, let A, and hence $(-A)$, have a dominant diagonal. Then workability of $(-A)$ (i.e. the possibility of positive net production in the context of Leontief systems) as defined in Nikaido (1968), 6.2, obtains. According to Theorem 6.3 of Nikaido (1968), $(-A)$ has a nonnegative inverse. As a consequence, $\frac{ds^*}{dt}(\bar{t})$, which equals $(-A)^{-1}$ applied to the positive vector $(\partial_t \partial_i \Pi_i)$, is strictly positive. □

Remark 2. As pointed out in the introduction, the largest and the smallest equilibrium are locally increasing in t in a smooth supermodular game with $\partial_t \partial_i \Pi_i > 0$ for all $i = 1, \cdots, n$. If these equilibria lie in the interior of the strategy space, they have to satisfy the dominant diagonal property.

Remark 3. Theorem 1 is related to proposition 2.9 in Corchón (1996). This proposition says that equilibrium prices increase, if $\partial_t \partial_i \Pi_i > 0$ and the dominant diagonal condition holds.

One of several conditions equivalent to the workability of $(-A)$ states that $(-A)$ has all principal minors positive, i.e. $(-A)$ is a P-matrix. According to a theorem by Gale and Nikaido, $(-A)$ is a P-matrix if and only if $(-A)$ does not reverse (component by component) the sign of any vector different from 0 [cf. Nikaido (1968), Theorem 20.2]. It is apparent that, under the assumption of weak local supermodularity, local comparative statics of Nash equilibria work intrinsically by the same mathematical principles as the corresponding theory of linear economic models.

Observe that a P-matrix with dominant diagonal is positive quasi-definite. In particular, every eigenvalue of a P-matrix must have a positive real part. Clearly, this property is much stronger than requiring the matrix to have a positive determinant.

Now consider the vector field $(-\partial_i \Pi_i(s; \bar{t}))_{i=1,\cdots,n}$. We have just seen that the zero s^* of this vector field has index $+1$ if $(-\partial_j \partial_i \Pi_i(s^*; \bar{t}))_{i,j=1,\cdots,n}$ has a dominant diagonal [for vector fields and Euler numbers see, e.g., Milnor (1965), 6].

Corollary. Under the conditions of Theorem 1, the equilibrium $s^*(t)$ is strictly increasing only if the vector field $-(\partial_i \Pi_i(\cdot\,; \bar{t}))_{i=1,\cdots,n}$ has index $+1$ at s^*.

Since the space S of strategy profiles is convex, its Euler characteristic equals 1 and the Poincaré-Hopf theorem can easily be invoked to formulate a uniqueness result. Recall, though, that Gale and Nikaido have been able to derive uniqueness based on the concept of a P-matrix without using such topological tools [see Nikaido (1968), 20]. This illustrates the fact that a setting governed by P-matrices has a very particular structure.

It is generally considered natural to work in a framework of supermodularity in order to compare Nash equilibria. However, it is apparent from Theorem 1 and the subsequent discussion that the conditions entailing that a higher degree t of heterogeneity is associated with higher equilibrium prices are more restrictive than one might anticipate from Perloff's and Salop's analysis of the symmetric case. The crucial role of the assumption that marginal profits increase with product differentiation has been illustrated in the examples contained in Section 3, where an economic interpretation has also been given. The mathematical implications of the dominant diagonal assumption are clear, but it seems desirable to understand its economic interpretation. This issue will be addressed in the following subsection.

4.2 The case of constant marginal costs

Theorem 1 above, that highlights the role of the dominant diagonal property for comparative statics in the presence of strategic complementarity, has been formulated in a rather abstract setting. There has been no need to be specific with respect to payoffs or with respect to strategies. Such an approach is, of course, unsatisfactory from an economic perspective, since the dominant diagonal property has not been expressed in terms of the fundamental data of the model. Thus, we may ask in which situations the dominant diagonal condition follows from a more basic assumption directly imposed on demand. For that purpose, payoffs as well as strategies need to be specified in more precise economic terms.

To associate an economic meaning with the dominant diagonal property we assume now that each firm i has constant unit costs c_i. Then profits become $(p_i - c_i)f_i(p_i, p_{-i})$, where f_i denotes the demand for brand i. Since we want to impose assumptions directly on demand rather than on profits, we consider log-profits $\log(p_i - c_i) + \log(f_i(p_i, p_{-i}))$ as payoffs in order to separate demand additively from the mark-up term.

Next we specify the algebraic structure on the strategy space in order to describe a situation in which the dominant diagonal property holds. In case of perfectly divisible products it turns out to be appropriate for economic reasons to let the strategy space S_i consist of log-prices. In other words, the multiplicative structure on prices should be used.

Observe that an arbitrary real-valued function f defined on price systems p satisfies the supermodularity condition $\frac{\partial^2 f}{\partial p_j \partial p_i}(p_1, \cdots, p_n) \geq 0$ if and only if $\frac{\partial^2 f}{\partial s_j \partial s_i}(\varphi_1(s_1), \cdots, \varphi_n(s_n)) \geq 0$ for transformations φ_i with positive derivative.

A similar remark applies to the assumption that marginal payoffs increase with product differentiation. More precisely, if $\varphi_i' > 0$ then the inequality $\frac{\partial^2 \Pi_i}{\partial t \partial p_i}(p_1, \cdots, p_n; \bar{t}) > 0$ holds iff $\frac{\partial^2 \Pi_i}{\partial t \partial s_i}(\varphi_1(s_1), \cdots, \varphi_n(s_n); \bar{t}) > 0$. Therefore, in contrast to payoffs, strategies can be transformed freely in any way that is convenient to interpret the dominant diagonal property.

Since log-profits are taken as payoffs, we can express all assumptions in terms of the demand faced by a firm. The first assumption, which is often made in the context of Bertrand competition in order to ensure strategic complementarity, is fulfilled in examples commonly used in the context of price competition and constant marginal costs, [cf. Milgrom and Roberts (1990), p. 1271, or Vives (1990), p. 318]. Put $\pi = \log(p)$ and let d_i denote firm i's demand as a function of π and the parameter t. The assumption states that the elasticity of firm i's demand (which is, of course, negative) does not decrease if another firm raises its price.

(A-1) Weak cross monotonicity of demand elasticities:
$\partial_j \partial_i \log d_i(\pi; t) \geq 0$ for all $i = 1, \cdots, n$ and for all $j \neq i$.

Next we formulate the assumption of locally increasing marginal payoffs, that has been motivated and extensively interpreted in the context of the examples presented in Section 3, in terms of elasticities of demand. This assumption needs to hold at Nash equilibria only. Therefore, let $\pi^*(t)$ denote a Nash equilibrium of the economy with degree t of product differentiation.

(A-2) Demand elasticities increase with product differentiation:
$\partial_t \partial_i \log d_i(\pi^*(t); t) > 0$ for all i.

Our final assumption entails the uniqueness of Nash equilibrium. Formally, it amounts to a dominant diagonal assumption with equal weights. This assumption has a clear economic interpretation in the context of the discrete choice model presented in Section 2.2. In this setting, the elasticity $\partial_i \log d_i(\pi)$ of firm i's demand is composed of the following two effects:

i) firm i's market share declines, if i unilaterally raises its price;
ii) *conditional demand* changes, since the customers tend to substitute good 0 for brand i.

In contrast to (A-1), the dominant diagonal assumption focuses on effect ii). Since π denotes log-prices rather than prices themselves, the price ratio of any two brands remains unchanged if the derivative $\sum_{j=1}^{n} \partial_j d_i(\pi)$ is taken. As a consequence, firm i has the same group of customers before and after adding the amount $\Delta\pi > 0$ to each π_i. Therefore, $d_i(\pi_1 + \Delta\pi, \cdots, \pi_n + \Delta\pi) - d_i(\pi_1, \cdots, \pi_n)$ measures the change in demand induced by the price increase $\Delta\pi$ if the set of customers is kept fixed. For short, we call $\sum_{j=1}^{n} \partial_j \log d_i(\pi)$ the *elasticity of the conditional demand* for brand i, the condition requiring customers to stay loyal to the firm chosen before the price increase. The dominant diagonal assumption $\partial_i \sum_{j=1}^{n} \partial_j \log d_i(\pi) \leq 0$ states that the elasticity of firm i's conditional demand (which is negative) decreases with respect to i's price. Thus, the elasticity $\sum_{j=1}^{n} \partial_j \log d_i$ of conditional demand changes in the same direction as the elasticity of unconditional demand if firm i raises its price, although, according to (A-1), the cross effects $\partial_j \partial_i \log d_i, i \neq j$, work in the opposite direction.

(A-3) Weak monotonicity of the elasticity of conditional demand:
$\partial_i \sum_{j=1}^n \partial_j \log d_i(\pi; t) \leq 0 \quad$ for all i .

We are now ready to formulate an extension of Perloff's and Salop's result that does not rely on symmetry and that does also incorporate the issues of existence and uniqueness of Nash equilibria. It is assumed throughout that all demand and cost functions are twice continuously differentiable. Moreover, demand is positive so that the logarithm of the profit function is well defined.

Theorem 2. *Assume each firm $i = 1, \cdots, n$ has constant marginal costs $c_i > 0$ and the demand for each product is everywhere positive. Let i's strategy space be $S_i =]\log c_i, \infty[$ and let $\pi \in S = \prod_{i=1}^n S_i$, $t \in]t_l, t_h[$.*

a) *Assume there is, for each $t \in]t_l, t_h[$, a strategy profile $\tilde{\pi}^t \in S$ such that there exists, for every firm i, an optimal response to $\tilde{\pi}^t_{-i}$ and any optimal response to $\tilde{\pi}^t_{-i}$ lies strictly below $\tilde{\pi}^t_i$. Moreover, let assumption (A-1) hold for all $\pi \in S, t \in]t_l, t_h[$. Then there exists, for each $t \in]t_l, t_h[$, a Nash equilibrium $\pi^*(t) < \tilde{\pi}^t$.*
b) *Furthermore, if (A-1) and (A-3) hold for all $\pi \in S, t \in]t_l, t_h[$, then $\pi^*(t)$ is the unique Nash equilibrium for given $t \in]t_l, t_h[$.*
c) *If, in addition, (A-2) holds for $\pi^*(t), t \in]t_l, t_h[$, then $\pi^*(t)$ is strictly increasing in t for $t \in]t_l, t_h[$.*

Assumptions (A-1) and (A-3) are satisfied in the examples presented in Section 3. If d_i denotes demand in the CES case, we have $\partial_j \partial_i \log d_i(\pi) \geq 0$ and, moreover, $\sum_{j=1}^2 \partial_i \partial_j \log d_i(\pi) = 0$. Note that the latter fact implies the strict inequality $\sum_{j=1}^2 \partial_i \partial_j \log \Pi_i(\pi_j, t) < 0$. As pointed out in Section 3, assumption (A-2) on increasing elasticities of demand only holds in the examples, if t exceeds some critical level.

Proof. Since i's profit $\Pi_i(p; t) = (p_i - c_i) f_i(p; t)$ vanishes for $p_i = c_i$ and is positive for $p_i > c_i$, there is $\delta_i^t > 0$ such that no price $p_i \in [c_i, c_i + \delta_i^t]$ can be an optimal response to any $p_{-i} \in [c_{-i}, \exp(\tilde{\pi}^t_{-i})]$. We want to show that the boundary assumption stated in a) allows us to restrict strategy profiles to the complete lattice $\prod_{i=1}^n [\log(c_i + \delta_i^t), \tilde{\pi}_i^t]$.

To see this, consider log-profits $\Phi_i(\pi; t) = \log(e^{\pi_i} - c_i) + \log(d_i(\pi; t))$. Observe that (A-1) implies $\partial_j \partial_i \Phi_i(\pi; t) \geq 0$ for all π. Let $\pi_{-i} \leq \tilde{\pi}^t_{-i}$ and suppose $\pi_i' \in \operatorname{argmax} \Phi_i(\cdot, \pi_{-i}; t)$ and $\pi_i'' \in \operatorname{argmax} \Phi_i(\cdot, \tilde{\pi}^t_{-i}; t)$. From Topkis' monotonicity theorem we conclude $\max\{\pi_i', \pi_i''\} \in \operatorname{argmax} \Phi_i(\cdot, \tilde{\pi}^t_{-i}; t)$. The boundary assumption then yields $\max\{\pi_i', \pi_i''\} \leq \tilde{\pi}_i^t$. Therefore, $\Phi_i(\cdot, \pi_{-i}; t)$ must take its maximum in the compact interval $[\log(c_i + \delta_i^t), \tilde{\pi}_i^t]$.

Since the best responses of every player i to any $\pi_{-i} \leq \tilde{\pi}^t_{-i}$ are below $\tilde{\pi}_i^t$, any fixed point of the best reply correspondence of the game restricted to $\prod_{i=1}^n [\log(c_i + \delta_i^t), \tilde{\pi}_i^t]$ is a Nash equilibrium of the original game. By Topkis' monotonicity theorem, the largest (as well as the smallest) best response is increasing. The existence of a Nash equilibrium now follows from Tarski's fixed point theorem [see Section 2.1]. For every $t \in]t_l, t_h[$, let $\pi(t)$ be such a Nash equilibrium.

The uniqueness claimed in part b) is implied by a global version of the argument underlying Theorem 1, where the matrix $(-A)^{-1}$ maps the orthant $\mathbb{R}^n_+$ into itself. The basic tool is a lemma that follows from the mean value theorem by an elementary calculation [see Milgrom and Roberts (1990), p. 1272 and E. Dierker and Grodal (1996), p. 157].

Lemma. *Let t be given and consider two strategy profiles π' and π'' in S and let (A-1) and (A-3) hold along the line segment between π' and π''. If $\partial_i \Phi_i(\pi'; t) \geq \partial_i \Phi_i(\pi''; t)$ for $i = 1, \cdots, n$, then $\pi'' \geq \pi'$.*

Let t be given and consider any fixed point π of the best reply correspondence. Clearly, $\log(c_i + \delta_i^t) < \pi_i < \tilde{\pi}_i^t$ and we have $\partial_i \Phi_i(\pi; t) = 0$.

Now let π' and π'' be two fixed points. We have just shown that $\partial_i \Phi_i(\pi'; t) = \partial_i \Phi_i(\pi''; t) = 0$. Also, assumption (A-3) implies $\sum_{j=1}^n \partial_j \partial_i \Pi_i(\pi; t) < 0$ by definition of Φ. Thus, the lemma can be applied and we get uniqueness, since $\pi' \geq \pi''$ as well as $\pi' \leq \pi''$.

To show c), let $t' < t''$ be in $]t_l, t_h[$. Then, by (A-2), $\partial_i \Phi_i(\pi(t'); t'') > \partial_i \Phi_i(\pi(t'); t') = 0$ for all i if t' and t'' are sufficiently close, since $\pi(t')$ is defined to be the equilibrium for degree t' of product differentiation. By continuity, there exists a strictly positive vector a such that $\partial_i \Phi_i(\pi(t') + a; t'') > 0 = \partial_i \Phi_i(\pi(t''); t'')$ for all i. The above lemma yields $\pi(t'') \geq \pi(t') + a$. □

5. Strategic complementarity and heterogeneity

The third example studied in Section 3 relies upon increasing returns to scale. This phenomenon is of great importance for the oligopolistic markets under consideration as has, e.g., been pointed out by Anderson et al. (1992) [cf. Section 1]. We have seen in Section 3 that the existence of a Nash equilibrium in the presence of increasing returns depends on the degree of product heterogeneity. Thus, we are led to ask whether a sufficiently large degree of product differentiation helps to extend the lattice theoretical approach towards existence beyond the setting of Theorem 2. More precisely, we assume that demand functions satisfy assumption (A-1) and cost functions are concave and state a condition characterizing increasing best responses. For convenience, we drop the parameter t unless we want to vary the degree of product heterogeneity explicitly.

Proposition. *Assume $\partial_i d_i(\pi) < 0$, $\partial_j d_i(\pi) > 0$, and $\partial_j \partial_i \log d_i(\pi) > 0$. Moreover, suppose that firm i's cost function C_i satisfies $C_i' > 0$ and that the first order condition $\partial_i \Pi_i(\pi) = 0$ holds. Then the sign of $\partial_j \partial_i \Pi_i(\pi), i \neq j$, equals the sign of*

$$\left(1 - \frac{d_i(\pi)}{\partial_i d_i(\pi)} \frac{\partial_i \partial_j d_i(\pi)}{\partial_j d_i(\pi)}\right) - \frac{C_i''(d_i(\pi))}{C_i'(d_i(\pi))} d_i(\pi) \cdot (1 + \varepsilon_i(\pi)), \tag{13}$$

where $\varepsilon_i(\pi) = \partial_i d_i(\pi)/d_i(\pi)$ *is the elasticity of the demand for brand* i.

Proof. Firm i's profit is $\Pi_i(\pi) = e^{\pi_i} d_i(\pi) - C_i(d_i(\pi))$ and the first order condition $\partial_i \Pi_i(\pi) = 0$ amounts to $e^{\pi_i}(d_i(\pi) + \partial_i d_i(\pi)) = C_i'(d_i(\pi)) \cdot \partial_i d_i(\pi)$. Hence, $\partial_j \partial_i \Pi_i(\pi) = e^{\pi_i}[\partial_j d_i(\pi) + \partial_j \partial_i d_i(\pi)] - C_i'(x_i) \cdot \partial_j \partial_i d_i(\pi) - C_i''(x_i) \cdot \partial_j d_i(\pi) \cdot \partial_i d_i(\pi)$ where $x_i = d_i(\pi)$. We divide this equation by the positive number $(e^{\pi_i} \partial_j d_i(\pi))$ and use the above first order condition to obtain the conclusion. □

Before we discuss the implications of this condition for economies with concave cost functions we want to remark that the Proposition also shows the following well-known fact.

Remark 4. Assume $\partial_i d_i(\pi) < 0$, $\partial_j d_i(\pi) > 0$, and $\partial_j \partial_i \log d_i(\pi) > 0$. Then firm j exhibits monotone best responses, if its cost function C_i is convex, i.e. $C_i' \geq 0, C_i'' \geq 0$.

Assume now that C_i is concave and observe that $\varepsilon_i(\pi) < -1$ at an optimal response. Furthermore, note that the assumption $\partial_j \partial_i \log d_i(\pi) > 0$ is equivalent to

$$A = 1 - \frac{d_i(\pi)}{\partial_i d_i(\pi)} \frac{\partial_i \partial_j d_i(\pi)}{\partial_j d_i(\pi)} > 0. \tag{14}$$

The proposition states that an increasing best response at π requires the absolute value $|C_i''(x_i)\, x_i|/C_i'(x_i)$ of the elasticity of firm i's marginal costs to not exceed $A/|1 + \varepsilon_i(\pi)|$.

The demand function can be written as a product $d_i(\pi; t) = h_i(\pi_i; t) s_i(\pi; t)$, where the factor s_i represents the market share of firm i. In the examples of Section 3 we have put $h_i(\pi_i; t) = 1/e^{\pi_i}$ (Cobb-Douglas), but it is instructive to allow the elasticity of h_i to take values below -1. This case is of particular relevance, because a monopolist serving the market in the absence of product differentiation must have demand elasticity below -1 at an optimal price.

Putting $d_i(\pi; t) = h_i(\pi_i; t) s_i(\pi; t)$ we obtain for $j \neq i$:

$$\begin{aligned} \partial_i d_i(\pi; t) &= h_i(\pi_i; t) \partial_i s_i(\pi; t) + \partial_1 h_i(\pi_i; t) s_i(\pi; t), \\ \partial_j d_i(\pi; t) &= h_i(\pi_i; t) \partial_j s_i(\pi; t), \\ \partial_i \partial_j d_i(\pi; t) &= h_i(\pi_i; t) \partial_i \partial_j s_i(\pi; t) + \partial_1 h_i(\pi_i; t) \partial_j s_i(\pi; t). \end{aligned} \tag{15}$$

Hence we have, omitting the argument $(\pi; t)$,

$$\begin{aligned} \frac{d_i}{\partial_i d_i} \cdot \frac{\partial_i \partial_j d_i}{\partial_j d_i} &= \frac{s_i}{h_i \cdot \partial_i s_i + \partial_1 h_i \cdot s_i} \cdot \frac{h_i \cdot \partial_i \partial_j s_i + \partial_1 h_i \cdot \partial_j s_i}{\partial_j s_i} \\ &= \left(\frac{\partial_i s_i}{s_i} + \frac{\partial_1 h_i}{h_i}\right)^{-1} \cdot \left(\frac{\partial_i \partial_j s_i}{\partial_j s_i} + \frac{\partial_1 h_i}{h_i}\right). \end{aligned} \tag{16}$$

Furthermore, in the discrete choice model described in Section 2.2, the identity $s_i(\pi;t) = s_i(\pi/t;1)$ holds by definition of the parameter t. Therefore, $\partial_j s_i(\pi;t) = \partial_j s_i(\pi/t;1)/t$ and $\partial_i\partial_j s_i(\pi;t) = \partial_i\partial_j s_i(\pi/t;1)/t^2$. For all π and $i = 1,\cdots,n$, let $|\partial_i s_i(\pi;1)/s_i(\pi;1)| \leq k_i$ and $|\partial_i\partial_j s_i(\pi;1)/\partial_j s_i(\pi;1)| \leq K_i$. Then the inequalities

$$\left|\frac{\partial_i s_i(\pi;t)}{s_i(\pi;t)}\right| \leq \frac{k_i}{t} \quad \text{and} \quad \left|\frac{\partial_i\partial_j s_i(\pi;t)}{\partial_j s_i(\pi;t)}\right| \leq \frac{K_i}{t} \tag{17}$$

hold uniformly in π. It follows that (16) converges to 1 if t approaches ∞, since $\varepsilon_i(\pi;t) = \partial_i s_i(\pi;t)/s_i(\pi;t) + \partial_1 h_i(\pi_i;t)/h_i(\pi_i;t) < \partial_1 h_i(\pi_i;t)/h_i(\pi_i;t) < -1$ at an optimal response of firm i. Note that $1+\varepsilon_i(\pi;t)$ tends to 0 in Example 3 in Section 3.

Remark 5. The expression A in (14) tends to 0 if t approaches ∞. Moreover, $\partial_j\partial_i\Pi_i(\pi;t)$ becomes negative, if i's cost function satisfies $C'' < 0$ and $1 + \varepsilon_i(\pi;t)$ does not tend to zero.

6. Quasiconcavity and heterogeneity

The proposition in Section 5 is unsatisfactory for the following reasons. First, the crucial term A in (13) reflecting the assumption that the firm's demand satisfies (A-1) does not stay bounded away from 0. Therefore, assumption (A-1) loses its bite for large t. Furthermore, Grandmont (1992) has shown that a sufficiently strong heterogeneity of tastes entails that the salient properties of Cobb-Douglas demand functions hold approximately. Since heterogeneity is modeled here in a way closely related to his and $\varepsilon(\pi;t) = -1$ for Cobb-Douglas demand functions, one cannot expect the term $1 + \varepsilon(\pi;t)$ to stay away from 0 for very large t. But then the proposition in Section 5 leaves us with the task to determine the sign of a limit of the type "0 minus 0". This exercise, however, is of rather limited interest, since a profit maximizing firm facing a demand that is approximately Cobb-Douglas sells nearly nothing at a price close to infinity. Clearly, a statement for a fixed, finite t is much more desirable than a limit result of the type just described, if one wants to show the existence of a Nash equilibrium.

Thus, we are led to explore to what extent heterogeneity helps to restore the quasiconcavity of the profit functions in oligopoly games with strictly concave cost functions. Indeed, numerical examples such as Example 3 in Section 3 show that a sufficient degree of heterogeneity leads to quasiconcavity. This observation is fully in line with results in de Palma et al. (1985) on product differentiation.

Theorem 3 below states that firm i's profit function is quasiconcave if the absolute value of the elasticity of i's marginal cost function is less than $1/|\varepsilon_i(\pi;t)|$ provided the demand function is logconcave. This result holds for every degree t of product differentiation. As before, we consider log-prices

π as the underlying variables. Logconcavity of the demand $d(\pi)$ means that $\log d$ is concave or, in other words, that the elasticity $\partial_i d(\pi)/d(\pi)$ of demand is decreasing. Therefore, logconcavity fits very well to the assumptions made in Section 4.

Theorem 3. *Assume that $\partial_i d_i(\pi) < 0$ and that $\log d_i(\pi)$ is concave with respect to π_i, i.e. $d_i(\pi) \cdot \partial_i \partial_i d_i(\pi) < (\partial_i d_i(\pi))^2$. Moreover, suppose that firm i's cost function C_i is concave, satisfies $C_i' > 0$, and the following condition: For any π such that $\partial_i \Pi_i(\pi) = 0$, the absolute value of the elasticity of i's marginal costs does not exceed that of the inverse of its elasticity of demand, that is to say, $|C_i''(x_i)\, x_i|/C_i'(x_i) \leq 1/|\varepsilon_i(\pi)|$, where $x_i = d_i(\pi)$ and $\varepsilon_i(\pi) = \partial_i d_i(\pi)/d_i(\pi)$. Then firm i's profit function $\Pi_i(\pi) = e^{\pi_i} d_i(\pi) - C_i(d_i(\pi))$ is quasiconcave with respect to π_i.*

Proof. It is sufficient to show $\partial_i \partial_i \Pi_i(\pi) < 0$ for any π satisfying the first order condition $\partial_i \Pi_i(\pi) = 0$ for profit maximization. In this case we have $e^{\pi_i} = C_i'(d_i(\pi)) \cdot \partial_i d_i(\pi)/(\partial_i d_i(\pi) + d_i(\pi))$. Observe that the denominator is negative. Writing for short $x_i = d_i(\pi)$ and substituting the expression for C_i' as given by the first order condition, the second order partial derivative $\partial_i \partial_i \Pi_i(\pi)$ obtains as

$$\frac{e^{\pi_i}}{\partial_i d_i(\pi)} \{2(\partial_i d_i(\pi))^2 - x_i \cdot \partial_i \partial_i d_i(\pi) + x_i \cdot \partial_i d_i(\pi) - \frac{C_i''(x_i)(\partial_i d_i(\pi))^3}{e^{\pi_i}}\}. \tag{18}$$

By assumption $(\partial_i d_i(\pi))^2 - x_i \cdot \partial_i \partial_i d_i(\pi) > 0$. Thus $\partial_i \partial_i \Pi_i(\pi)$ is negative, if

$$\begin{aligned} (\partial_i d_i(\pi))^2 + x_i \cdot \partial_i d_i(\pi) - \frac{C_i''(x_i) \cdot (\partial_i d_i(\pi))^3 \cdot (x_i + \partial_i d_i(\pi))}{C_i'(x_i) \cdot \partial_i (d_i(\pi))} &= \\ \partial_i d_i(\pi) \cdot (\partial_i d_i(\pi) + x_i)\{1 + \frac{|C_i''(x_i)| \cdot \partial_i d_i(\pi)}{C_i'(x_i)}\} &\geq 0. \end{aligned} \tag{19}$$

The last inequality, however, holds due to the assumption $|C_i''(x_i)\, x_i|/C_i'(x_i) \leq 1/|\varepsilon_i(\pi)|$. □

This proof also shows the following fact:

Remark 6. The profit function Π_i is quasiconcave with respect to π_i, if $\log d_i(\pi)$ is concave with respect to π_i and C_i is convex, i.e. $C_i' \geq 0, C_i'' \geq 0$.

Corollary. Let $\log d_i(\pi)$ be concave with respect to π_i and let the cost function C_i be isoelastic and concave, i.e. $C_i(x_i) = \gamma_i x_i^\alpha$ for $0 < a < 1$. Then the profit function Π_i is quasiconcave with respect to π_i, if the elasticity $\varepsilon_i(\pi) = \partial_i d_i(\pi)/d_i(\pi)$ of firm i's demand satisfies $|\varepsilon_i(\pi)| \leq 1/(1-\alpha)$ for all π such that $\partial_i \Pi_i(\pi) = 0$. □

Proof.

$$\frac{|C_i''(d_i(\pi))|}{C_i'(d_i(\pi))} d_i(\pi) = \frac{(1-\alpha)\gamma_i (d_i(\pi))^{\alpha-2}}{\gamma_i (d_i(\pi))^{\alpha-1}} d_i(\pi) = 1 - \alpha\,. \tag{20}$$

Example: Consider the demand function $d_1(\pi_1, \pi_2) = (e^{\pi_1})^{-r-1}/(e^{-r\pi_1} + e^{-r\pi_2})$ underlying example 3 in Section 3 and remember that the degree of product differentiation is given by $t = 1/r$. A short calculation yields that $\log d_1$ is concave with respect to π_1 and that $\varepsilon_1(\pi_1, \pi_2) = -1 - [t(1 + \exp{(\pi_2 - \pi_1)/t})]^{-1} > -1 - 1/t$. Clearly, for any elasticity $\alpha \in]0, 1[$ the inequality $|\varepsilon_1(\pi)| \leq 1/(1-\alpha)$ is satisfied uniformly in π, if the degree of product differentiation t becomes sufficiently large. In particular, for any $0 < \alpha < 1$, the profit function $\Pi_1(\pi)$ of firm 1 is quasiconcave with respect to π_1 if $t \geq (1 - \alpha)/\alpha$. In Example 3 above the elasticity of the cost function of both firms has been specified as 0.9. According to Brouwer's fixed point theorem there exists a Bertrand-Nash equilibrium for $t \geq 1/9$. To get a feeling for the degree of product differentiation at $t = 1/9$ observe that, expressed in terms of market power in a symmetric duopoly, this number amounts to a markup of about 22%. Actually, since we have disregarded the term $1 + \exp{(\pi_2 - \pi_1)/t}$ in the formula for $\varepsilon_1(\pi_1, \pi_2)$, the bound 1/9 is not tight.

7. Conclusions

In this paper we have analyzed the bearing of product heterogeneity on markets with price competition. The intuition underlying the comparison of equilibria has been expressed by Anderson et al. (1992), p. 186 f., as follows: "As preference intensity rises, consumer tastes become increasingly different, and each variant has more and more "loyal" consumers prepared to buy it even at a premium. This means that demands become more inelastic and firms charge higher prices." Our analysis has shown that this statement is incomplete, since it only refers to the market power all firms together exert over the consumers. What is missing are considerations of how the strategic position of an individual firm in relation to its competitors is affected by the degree of product differentiation.

Two effects must be controlled in order to obtain an incentive for a firm to increase its price if the degree t of product differentiation rises. First, a firm may lose rather than gain market power, since an a priori existing competitive advantage, which might be caused by lower unit costs, is weakened if products become more differentiated. In this case the premia consumers are willing to pay are *decreasing* with respect to heterogeneity and thus render firm specific demand *less elastic*. Second, the elasticity of demand is also influenced by a change in market share. Thus, market shares must also be sufficiently stable. The analysis of the examples in Section 3 shows that the crucial condition requiring marginal profits to increase with product heterogeneity can easily be violated.

We have introduced a model of horizontal product differentiation and posed the question of how market power and product differentiation are related within the theory of supermodular games. Even if marginal profits are increasing and strategic complementarity is assumed to hold at a particular equilibrium, a dominant diagonal condition is necessary and sufficient to ensure that the equilibrium prices grow if product differentiation increases. The dominant diagonal condition is needed to restrict the degree of supermodularity.

Theorem 2 deals with the case of constant marginal costs. Imposing assumptions on elasticities of demand functions only we derive the existence of a unique equilibrium and the absence of counterintuitive behavior. Therefore, Theorem 2 strengthens the result in Perloff and Salop (1985).

However, if marginal costs are decreasing, the situation becomes more complicated. The effects listed above, that may induce a firm to raise its price, have an additional impact on marginal profits and thereby on the comparative statics via the concavity of the cost function. Moreover, equilibria may fail to exist, since the concavity of the cost function destroys quasiconcavity of the profit function. Finally, strategic complementarity, which is considered to be quite plausible in the framework of price competition among firms with constant marginal costs, may be lost for the same reason.

Example 3 in Section 3 shows that the lack of supermodularity for small degrees t of product differentiation may vanish if t increases. However, it is argued in Section 5 that there is no clear-cut relation between a sufficiently large degree of heterogeneity, the degree of concavity of the cost function, and supermodularity of the profit function. By contrast, if supermodularity is replaced by quasiconcavity, the picture becomes much brighter. In Section 6 we have stated a sharp lower bound (depending on the elasticity of the marginal cost function) for the degree of product differentiation that entails quasiconcavity of the profit function provided demand is logconcave. Thus, in case of strictly decreasing marginal costs and logconcave demand, the existence of an equilibrium can be shown by Brouwer's fixed point theorem, if the cost function is not too concave.

Summarizing, the intuition that more product differentiation leads to higher market power rests on restrictive assumptions which often remain hidden. This is particularly true, if increasing returns to scale, which are used to explain why not every taste is catered to, are reflected by decreasing marginal costs rather than by fixed costs associated with constant marginal costs.

References

Anderson, S.P., de Palma, A., Thisse, J.-F. : Discrete Choice Theory of Product Differentiation. MIT Press, Cambridge, Massachusetts 1992

Anderson, S.P., de Palma, A.: From local to global competition. discussion paper, University of Virginia 1996

Corchón, L.: Theories of Imperfectly Competitive Markets. Lecture Notes in Economics and Mathematical Systems 442, Springer, Heidelberg 1996

de Palma, A., Ginsburgh, V., Papageorgiou, Y.Y., Thisse, J.-F.: The principle of minimal differentiation holds under sufficient heterogeneity. Econometrica **53**, 767-781 (1985)

Dierker, E., Grodal, B.: Profit maximization mitigates competition. Economic Theory **7**, 139-160 (1986)

Grandmont, J.-M.: Transformations of the commodity space, behavorial heterogeneity, and the aggregation problem. Journal of Economic Theory **57**, 1-35 (1992)

Milgrom P., Roberts, J.: Rationalizability, learning, and equilibrium in games with strategic complementarities. Econometrica **58**, 1255-1277 (1990)

Milgrom P., Shannon, C.: Monotone comparative statics. Econometrica **62**, 157-180 (1994)

Nikaido, H.: Convex Structures and Economic Theory. Academic Press, New York 1968

Perloff, J., Salop, S.: Equilibrium with product differentiation. Review of Economic Studies, LII, 107-120 (1985)

Tarski, A.: A lattice-theoretical fixpoint theorem and its applications. Pacific Journal of Mathematics **5**, 285-309 (1955)

Topkis, D.: Minimizing a submodular function on a lattice. Operations Research **26**, 305-321 (1978)

Topkis, D.: Equilibrium points in nonzero-sum n-person submodular games. Siam Journal of Control and Optimization **17**, 773-787 (1979)

Vives, X.: Nash equilibrium with strategic complementarities. Journal of Mathematical Economics **19**, 305-321 (1990)

Adv. Math. Econ. 1, 69–82 (1999)

Advances in
MATHEMATICAL
ECONOMICS

A Remark on default risk models

Shigeo Kusuoka

Graduate School of Mathematical Sciences, University of Tokyo, 3-8-1 Komaba, Meguro-ku, Tokyo 153-0041, Japan

Received: April 21, 1998

JEL classification: G12

Mathematics Subject Classification (1991): 60G44, 90A99

Summary. We study some mathematical models on default risk. First, we study a "standard model" which is an abstract setting widely used in parctice. Then we study how the hazard rates changes, if we change a basic probability measure. We show that the usual assumptions on hazard rates hold in a standard model, but do not hold in general if we change a basic measure. Finally we study a filtering model.

1. Introduction

Let $(\Omega, \mathcal{F}, \{\mathcal{F}_t\}_{t\in[0,\infty)}, P)$ be a complete probability space with filtration satisfying the usual hypothesis. Let τ be a stopping time, and let $N_t = 1_{\{t\geq\tau\}}$. Then N_t is a submartingale, and so by Doob-Meyer's Theorem, there is a unique predictable increasing process A_t such that $A_0 = 0$ and $N_t - A_t$ is a martingale. In Mathematical Finance, a stopping time τ is sometimes regarded as a default time, and often assume that A_t is absolutely continuous. Then there is a predictable process λ_t such that $A_t = \int_0^t (1 - N_s)\lambda_s ds$, $t > 0$. But also, assuming some more assumptions, sevral authors derived the following equation

$$P(\tau > s|\mathcal{F}_t) = (1 - N_t)E^P[\exp(-\int_t^s \lambda_u du)|\mathcal{F}_t], \qquad s > t, \tag{1.1}$$

or similar equations, which is practically useful to compute the price of defaultable securities (e.g. Jarrow-Turnbull [4], Duffie-Singleton [3], Duffie-Schroder-Skidas [2]).

In this paper, we show that the equation (1.1) holds in, what we call, a standard model, but does not hold in general if we take other equivalent probability measures. We show modified equations hold in general. Unfortunately we have to introduce a new measure as a reference measure to describe modified equations for each default time, and so our results may not be useful in practice. However, credit derivatives were introduced recently, and one has to think of the model including many default times. So models get more complicated. This paper is a warning that one has to be careful about the assumptions to use convinient formulas in default risk models.

2. Standard model

Let $(\Omega, \mathcal{F}, P)$ be a complete probability space . Let $T > 0$ and fix it throughout the paper. Let $\mathcal{N} = \{A \in \mathcal{F};\ P(A) = 0 \text{ or } 1\}$, $\{\mathcal{G}_t\}_{t\in[0,T]}$ be a filtration over $(\Omega, \mathcal{F}, P)$ and τ_k, $k = 1, \ldots, N$, be $[0, \infty)$ -valued random variables. We assume the following.
(A-1) $\mathcal{N} = \mathcal{G}_0$ and $\{\mathcal{G}_t\}_{t\in[0,T]}$ is a d-dimensional weakly Brownian filtration, where $0 \leq d \leq \infty$.
(A-2) The probability distributions of τ_k, $k = 1, \ldots, N$, are continuous, and $P(\tau_k = \tau_\ell) = 0$, $k \neq \ell$.

Here we say that $\{\mathcal{G}_t\}_{t\in[0,T]}$ is a d-dimensional weakly Brownian filtration, if there is a d-dimensional P - $\{\mathcal{G}_t\}_{t\in[0,T]}$-Brownian motion satisfying the following.
(R) For any P - $\{\mathcal{G}_t\}_{t\in[0,T]}$-square integrable martingale Z_t, there is a $\{\mathcal{G}_t\}_{t\in[0,T]}$-predictable process $f : [0, T] \times \Omega \to \mathbf{R}^d$ with $E^P[\int_0^T |f(t)|^2 dt] < \infty$ such that

$$Z_t = Z_0 + \int_0^t f(s) dB_s \qquad t \in [0, T].$$

We call such a Brownian motion $\{B_t\}_{t\in[0,T]}$ satisfying the condition (R) a Brownian base of $(\Omega, \{\mathcal{G}_t\}_{t\in[0,T]}, P)$.

Let $N_k(t) = 1_{\{t \geq \tau_k\}}$, $t \in [0, T]$, $k = 1, \ldots, N$, and let

$$\mathcal{F}_t = \mathcal{G}_t \vee \sigma\{\tau_k \wedge t;\ k = 1, \ldots, N\}, \qquad t \in [0, T].$$

We assume the following moeover.
(A-3) The filtration $\{\mathcal{F}_t\}_{t\in[0,T]}$ is right continuous.

Note that the assumptions (A-1), (A-2), (A-3) are stable if we take another equivalent probability measure $\tilde{P}$ as a basic measure instead of P.

Then one can easily see the following.

Proposition 2.1. *For any $\{\mathcal{F}_t\}_{t\in[0,T]}$ progressively measurable process f : $[0, T] \times \Omega \to \mathbf{R}$, there is a measurable function $g : [0, T] \times [0, T]^N \times \Omega \to \mathbf{R}$ such that*
$g(\cdot, s_1, \ldots, s_N, *) : [0, T] \times \Omega \to \mathbf{R}$, $s_1, \ldots, s_N \in [0, T]$, is $\{\mathcal{G}_t\}_{t\in[0,T]}$ progressively measurable and

$$f(t) = g(t, \tau_1 \wedge t, \ldots, \tau_N \wedge t) \qquad a.e. t \in [0, T], \quad P - a.s$$

Now we assume moreover the following.
(M-1) There exist $\{\mathcal{G}_t\}_{t\in[0,T]}$-progressively measurable processes $\lambda_k : [0,T) \times \Omega \to [0,T]$, $k = 1,\ldots,N$, such that

$$M_k(t) = N_k(t) - \int_0^t (1 - N_k(s))\lambda_k(s)ds$$

are P -$\{\mathcal{F}_t\}_{t\in[0,T]}$ -martingales.
(M-2) Any P - $\{\mathcal{G}_t\}_{t\in[0,T]}$-martingale is a P - $\{\mathcal{F}_t\}_{t\in[0,T]}$-martingale.
Let

$$X_k(t) = \exp(\int_0^t \lambda_k(s)ds)(1 - N_k(t)), \qquad t \in [0,T], \quad k = 1,\ldots,N.$$

Then we easily see the following.

Proposition 2.2. *$\{X_k(t)\}_{t\in[0,T]}$, $k = 1,\ldots,N$, are locally bounded P - $\{\mathcal{F}_t\}_{t\in[0,T]}$ -martingales. Moreover, $[X_k, X_\ell] = 0$, $k \neq \ell$, and*

$$X_k(t) = 1 - \int_0^t X_k(s-)dM_k(s), \qquad t \in [0,T].$$

Let us fix a Brownian base $\{B(t)\}_{t\in[0,T]}$ of $(\Omega, \{\mathcal{G}\}_{t\in[0,T]}, P)$.

Theorem 2.3. *For any P - $\{\mathcal{F}_t\}_{t\in[0,T]}$- square integrable martingale $\{Y_t\}_{t\in[0,T]}$, there are $\{\mathcal{F}_t\}_{t\in[0,T]}$- predictable processes $f : [0,T] \times \Omega \to \mathbf{R}^d$ and $\tilde{f}_k : [0,T] \times \Omega \to \mathbf{R}$, $k = 1,\ldots,N$, such that*

$$E^P[\int_0^T |f(t)|^2 dt] < \infty, \qquad E^P[\int_0^T |\tilde{f}_k(t)|^2 \lambda_k(t)dt] < \infty, \ k = 1,\ldots,N,$$

and

$$Y_t = Y_0 + \int_0^t f(s)dB(s) + \sum_{k=1}^N \int_0^t \tilde{f}_k(s)dM_k(s) \qquad t \in [0,T].$$

Proof. Step 1. Let $\mathcal{H}$ be a linear span of $Z\prod_{j=1}^r (1 - N_{k_j}(s_j))$, where Z is a bounded $\mathcal{G}_T$-measurable random variables and $0 \leq r \leq N$, $1 \leq k_1 < k_2 < \ldots < k_r \leq N$ and $0 \leq s_j \leq T$, $j = 1,\ldots,r$. It is obvious that $\mathcal{H}$ is an algebra. Note that

$$\tau_k \wedge T = \lim_{n\to\infty} \sum_{i=1}^n ((1 - N_k(\frac{i-1}{n}T)) - (1 - N_k(\frac{i}{n}T)))\frac{i}{n}T + (1 - N_k(T))T.$$

So we see that $\mathcal{H}$ is dense in $L^2(\Omega, \mathcal{F}_T, P)$.

Step 2. Let Z is a bounded $\mathcal{G}_T$-measurable random variables and $0 \leq r \leq N$, $1 \leq k_1 \leq k_2 \leq \ldots \leq k_r \leq N$ and $0 \leq s_j \leq T$, $j = 1,\ldots,r$. Let

$$Z' = Z \exp(-\sum_{i=1}^{r} \int_0^{s_i} \lambda_{k_i}(t)dt)$$

Then Z' is $\mathcal{G}_T$-measurable. Let $Z'_t = E^P[Z'|\mathcal{F}_t]$, $t \geq 0$. Then Z'_t is a P - $\{\mathcal{F}_t\}_{t\in[0,T)}$- continuous bounded martingale. So $[Z', X_k] = 0$, $k = 1, \ldots, N$. Note that

$$\begin{aligned} & Z \prod_{j=1}^{r} (1 - N_{k_j}(s_j)) \\ = \; & Z'_T \prod_{j=1}^{r} X_{k_j}(T \wedge s_j) \\ = \; & Z'_0 + \int_0^T \prod_{j=1}^{r} X_{k_j}(t \wedge s_j -) dZ'_t + \sum_{j=1}^{r} \int_0^{s_j} Z'_{t-} \prod_{i=1}^{r} X_{k_i}(t \wedge s_i -) dM_{k_j}(t) \end{aligned}$$

This implies our theorem. ∎

For each subset I of $\{1, \ldots, N\}$, let

$$\mathcal{F}_t^I = \mathcal{G}_t \vee \sigma\{\tau_k \wedge t;\ k \in I\} \qquad t \in [0, T].$$

Then $B(t)$ and $M_k(t)$, $k \in I$, are P-$\{\mathcal{F}_t^I\}_{t\in[0,T]}$ -martingales. Then we have the following by a similar proof to Theorem 2.3.

Theorem 2.4. *Let I be a subset of $\{1, \ldots, N\}$. For any P - $\{\mathcal{F}_t^I\}_{t\in[0,T]}$- square integrable martingale $\{Y_t\}_{t\in[0,T]}$, there are $\{\mathcal{F}_t^I\}_{t\in[0,T]}$- predictable processes $f : [0,T] \times \Omega \to \mathbf{R}^d$ and $\tilde{f}_k : [0,T] \times \Omega \to \mathbf{R}$, $k \in I$, such that*

$$E^P[\int_0^T |f(t)|^2 dt] < \infty, \qquad E^P[\int_0^T |\tilde{f}_k(t)|^2 \lambda_k(t) dt] < \infty, \ k \in I,$$

and

$$Y_t = Y_0 + \int_0^t f(s)dB(s) + \sum_{k\in I} \int_0^t \tilde{f}_k(s) dM_k(s)$$

As a Corollary to Theorem 2.4, we have the following.

Proposition 2.5. *The filtration $\{\mathcal{F}_t^I\}_{t\in[0,T]}$ is right continuous, for any subset I*
of $\{1, \ldots, N\}$.

3. Equivalent probability measures

Let Q be a probability measure on $(\Omega, \mathcal{F})$ equivalent to P. Let

$$\rho_t = E^P[\frac{dQ}{dP}|\mathcal{F}_t], \qquad t \in [0,T].$$

We may assume that ρ_t is a cadlag process. We assume the following furthermore.

(Q) $\{\log \rho_t\}_{t\in[0,T]}$ is locally bounded.

Then by Theorem 2.3 there are predictable processes $\beta : [0,T]\times\Omega \to \mathbf{R}^d$, $\kappa_k : (-1,\infty)\times\Omega \to \mathbf{R}$, $k=1,\dots,N$, such that

$$\rho_t = 1 + \int_0^t \rho_{s-}(\beta(s)dB(s) + \sum_{k=1}^N \kappa_k(s)dM_k(s)), \qquad t\in[0,T].$$

Proposition 3.1.

$$\tilde{B}(t) = B(t) - \int_0^t \beta(s)ds$$

is a d-dimensional Q - $\{\mathcal{F}_t\}_{t\in[0,T]}$ *Brownian motion, and*

$$\tilde{M}_k(t) = N_k(t) - \int_0^t (1-N_k(s))(1+\kappa_k(s))\lambda_k(s)ds, \qquad k=1,\dots,N$$

are Q - $\{\mathcal{F}_t\}_{t\in[0,T]}$ *martingales.*

Proof. One can easily see

$$d(\rho_t N_k(t)) = \rho_t(1+\kappa_k(t))\lambda_k(t)dt + N_k(t-)d\rho_t + \rho_{t-}dM_k(t)$$

and

$$d(\rho_t B(t)) = \rho_{t-}\beta(t)dt + B(t)d\rho_t + \rho_{t-}dB(t)$$

Note that $\{Z_t\}_{t\in[0,T]}$ is a Q-martingale, if $\{\rho_t Z_t\}_{t\in[0,T]}$ is a P-martingale. Therefore we have our assertion. ∎

Theorem 3.2. *For any* Q - $\{\mathcal{F}_t\}_{t\in[0,T]}$- *square integrable martingale* $\{Y_t\}_{t\in[0,T]}$, *there are* $\{\mathcal{F}_t\}_{t\in[0,T]}$- *predictable processes* $f : [0,T]\times\Omega\to\mathbf{R}^d$ *and* $\tilde{f}_k : [0,T]\times\Omega\to\mathbf{R}$, $k=1,\dots,N$, *such that*

$$E^Q[\int_0^T |f(t)|^2 dt] < \infty,$$

$$E^Q[\int_0^T |\tilde{f}_k(t)|^2(1+\kappa_k(t))\lambda_k(t)dt] < \infty, \ k=1,\dots,N,$$

and

$$Y_t = Y_0 + \int_0^t f(s)d\tilde{B}(s) + \sum_{k=1}^N \int_0^t \tilde{f}_k(s)d\tilde{M}_k(s), \qquad t\in[0,T].$$

Proof. We may assume that Y_t is bounded. Note that $\rho_t Y_t$ is a P-local square integrable martingale. Therefore

$$Y'_t = \int_0^t \rho_{s-}^{-1} d(\rho_s Y_s) - \int_0^t \rho_{s-}^{-1} Y_{s-} d\rho_s$$

is also so. By Theorem 2.3, there are predictable processes $f, \tilde{f}_k$ such that

$$Y'_t = \int_0^t f(s)dB(s) + \sum_{k=1}^N \int_0^t \tilde{f}_k(s)dM_k(s)$$

Since

$$dY_t = dY'_t - \rho_{t-}^{-1}[\rho, Y]_t = f(t)d\tilde{B}(t) + \sum_{k=1}^N \tilde{f}_k(t)d\tilde{M}_k(t)$$

we have our assertion. ∎

For any subset I of $\{1, \ldots, N\}$, let $\tau_I = \wedge_{k\in I}\tau_k = \min\{\tau_k;\ k \in I\}$.

Proposition 3.3. *Let I be a subset of $\{1, \ldots, N\}$. Then there are $\{\mathcal{F}_t^{I^c}\}_{t\in[0,T]}$ predictable processes $\beta^I : [0,T]\times\Omega :\to \mathbf{R}^d$ and $\kappa_k^I : [0,T]\times\Omega :\to (-1,\infty)$, $k = 1, \ldots, N$ such that*

$$\beta(t) = \beta^I(t), \qquad \kappa_\ell(t) = \kappa_k^I(t) \qquad a.e.t \in [0, T\wedge\tau_I] \quad P-a.s.$$

Here $I^c = \{1, \ldots, N\}\setminus I$.

Proof. We may assume that $I = \{1, \ldots, \ell\}$, $1 \le \ell \le N$. By Proposition 2.3 there is a $g : [0,T]\times[0,T]^N\times\Omega \to \mathbf{R}^d$ such that

$$\beta(t) = g(t, \tau_1\wedge t, \ldots, \tau_N\wedge t), \qquad a.e.t.$$

Let

$$\tilde{\beta}^I(t) = g(t, t, \ldots, t, \tau_{\ell+1}\wedge t, \ldots, \tau_N\wedge t),$$

and let

$$\beta^I(t) = \tan(\liminf_{h\downarrow 0} \frac{1}{h}\int_{0\vee(t-h)}^t \arctan(\tilde{\beta}^I(s))ds), \quad t \in [0,T].$$

Then we see that β^I is $\{\mathcal{F}_t^{I^c}\}_{t\in[0,T]}$ predictable and

$$\beta(t) = \beta^I(t), \qquad a.e.t \in [0, T\wedge\tau_I] \quad P-a.s.$$

The other are similar. ∎

Let us fix a subset I of $\{1, \ldots, N\}$. Let $\rho^I(t)$ be a solution of the following SDE

$$d\rho^I(t) = \rho^I(t-)(\beta^I(t)dB(t) + \sum_{k \in I^c} \kappa_k^I(t)dM_k(t)), \qquad \rho^I(0) = 1$$

We assume the following furthermore.
$(Q - I)$ $E^P[\rho^I(T)] = 1$.

Then we can define a probability measure Q^I on $(\Omega, \mathcal{F})$ by $dQ^I = \rho^I(T)dP$. Then we have the following.

Proposition 3.4. *For any bounded random variable Z, let*

$$Z_t^I = E^{Q^I}[Z|\mathcal{F}_t^{I^c}], \qquad t \in [0, T]$$

Then $Z_{t\wedge\tau_I}^I$ is a Q -$\{\mathcal{F}_t\}_{t\in[0,T]}$ martingale , and $[Z^I(\cdot \wedge \tau_I), N_k] = 0$, $k \in I$.

Proof. By applying Proposition 3.1 and Theorem 2.3 to Q^I we see that

$$B'(t) = B(t) - \int_0^t \beta^I(s)ds$$

is a Q^I - $\{\mathcal{F}_t^{I^c}\}_{t\in[0,T]}$ Brownian motion, and

$$M_k'(t) = N_k(t) - \int_0^t (1 - N_k(s))(1 + \kappa_k^I(s))ds, \qquad k \in I^c,$$

is a Q^I - $\{\mathcal{F}_t^{I^c}\}_{t\in[0,T]}$ martingale. Also, there are $\{\mathcal{F}_t^{I^c}\}_{t\in[0,T]}$- predictable processes $f : [0, T] \times \Omega \to \mathbf{R}^d$ and $\tilde{f}_k : [0, T] \times \Omega \to \mathbf{R}$, $k \in I^c$, such that

$$E^{Q^I}[\int_0^T |f(t)|^2 dt] < \infty,$$

$$E^{Q^I}[\int_0^T |\tilde{f}_k(t)|^2(1 + \kappa_k^I(t))\lambda_k(t)dt] < \infty, \quad k \in I^c,$$

and

$$Z_t = Z_0 + \int_0^t f(s)dB'(s) + \sum_{k\in I^c} \int_0^t \tilde{f}_k(s)dM_k'(s).$$

Note that

$$B'(t \wedge \tau_I) = \tilde{B}(t \wedge \tau_I), \qquad M_k'(t \wedge \tau_I) = \tilde{M}_k(t \wedge \tau_I), \quad k \in I^c.$$

So we have

$$Z_{t\wedge\tau_I} = Z_0 + \int_0^{t\wedge\tau_I} f(s)d\tilde{B}(s) + \sum_{k\in I^c} \int_0^{t\wedge\tau_I} \tilde{f}_k(s)d\tilde{M}_k(s).$$

This proves our assertion. ∎

Let $N_I(t) = 1_{\{t\geq\tau_I\}}$. Then we see that

$$\tilde{M}_I(t) = N_I(t) - \int_0^t (1 - N_I(s))(\sum_{k\in I}(1 + \kappa_k^I(s))\lambda_k(s))ds$$

is a Q -$\{\mathcal{F}_t\}_{t\in[0,T]}$ martingale . So we have the following.

Proposition 3.5. *For any bounded $\mathcal{F}_T^{I^c}$ measurable random variable Z,*

$$E^Q[Z(1-N_I(s))|\mathcal{F}_t]$$

$$= (1-N_I(t))E^{Q^I}[Z\exp(-\int_t^s(\sum_{k\in I}(1+\kappa_k^I(u))\lambda_k(u)du)|\mathcal{F}_t^{I^c}], \quad 0 \le t \le s \le T.$$

Proof. Let

$$Z' = Z\exp(-\int_0^s(\sum_{k\in I}(1+\kappa_k^I(u))\lambda_k(u)du).$$

Then Z' is $\mathcal{F}_T^{I^c}$ measurable. Let

$$Z'_t = E^{Q^I}[Z|\mathcal{F}_t^{I^c}],$$

and let

$$X'_t = (1-N_I(t\wedge s))\exp(\int_0^{t\wedge s}(\sum_{k\in I}(1+\kappa_k^I(u))\lambda_k(u)du), \qquad t\in[0,T].$$

Then $Z'_{\cdot\wedge\tau_I}$ and $\{X'_t\}_{t\in[0,T]}$ are Q -$\{\mathcal{F}_t\}_{t\in[0,T]}$ local martingales , and $[Z'_{\cdot\wedge\tau_I}, X'] = 0$. So one can see that $Z'_{\cdot\wedge\tau_I}X'$ is also a local martingale. However,

$$Z'_{t\wedge\tau_I}X'_t = Z'_tX_t$$

So we see that $Z'_tX'_t$ is a bounded Q -$\{\mathcal{F}_t\}_{t\in[0,T]}$ martingale. Therefore

$$E^Q[Z(1-N_I(s))|\mathcal{F}_t] = E^Q[Z'_TX'_T|\mathcal{F}_t] = Z'_tX'_t.$$

This proves our assertion. ∎

Corollary 3.6. *For $t, s\in[0,T]$ with $t\le s$*

$$Q[\tau_I > s|\mathcal{F}_t] = (1-N_I(t))E^{Q^I}[\exp(-\int_t^s(\sum_{k\in I}(1+\kappa_k^I(u))\lambda_k(u)du)|\mathcal{F}_t^{I^c}].$$

Proposition 3.7. *Let $\{Z_t\}_{t\in[0,T]}$ be a bounded $\{\mathcal{F}_t^{I^c}\}_{t\in[0,T]}$ predictable process. Then*

$$E^Q[\int_t^T Z_s dN_I(s)|\mathcal{F}_t]$$

$$= (1-N_I(t))E^{Q^I}[\int_t^T Z_s(\sum_{k\in I}(1+\kappa_k^I(s))\lambda_k(s))$$

$$\exp(-\int_t^s(\sum_{k\in I}(1+\kappa_k^I(u))\lambda_k(u)du)|\mathcal{F}_t^{I^c}]$$

for any $t\in[0,T]$.

Proof. This follows from Proposition 3.5 by observing that

$$\int_t^T Z_s dN_I(s) = \int_t^T Z_s dM_I(s) + \int_t^T Z_s(1-N_I(s-))(\sum_{k\in I}(1+\kappa_k^I(s))\lambda_k(s))ds.$$

For the original measure P we have the following. This is the reason why we call it a standard model.

Corollary 3.8. *For any bounded $\mathcal{G}_T$ measurable random variable Z,*

$$E^P[Z(1-N_I(s))|\mathcal{F}_t]$$

$$= (1-N_I(t))E^P[Z\exp(-\int_t^s(\sum_{k\in I}\lambda_k(u)du)|\mathcal{F}_t], \qquad 0 \le t \le s \le T$$

In particular,

$$P[\tau_I > s|\mathcal{F}_t] = (1-N_I(t))E^P[\exp(-\int_t^s(\sum_{k\in I}\lambda_k(u)du)|\mathcal{F}_t].$$

for $t, s \in [0,T]$ with $t \le s$.

Proof. If we apply the results to the case $Q = P$, we have $Q^I = P$. Moreover, by the assumption (M-2), we see that

$$E^P[Z|\mathcal{G}_t] = E^P[Z|\mathcal{F}_t], \qquad t \in [0,T]$$

for any bounded $\mathcal{G}_T$ measurable random variable Z. So we have our assertion from Proposition 3.5 and Corollary 3.6. ∎

Example. Let $T > 0$, $\lambda_1, \lambda_2 > 0$, $\Omega = [0,\infty)^2$, $\mathcal{F}$ be a Borel algebra over Ω,

$$P(dx, dy) = (\lambda_1\lambda_2)\exp(-\lambda_1 x - \lambda_2 y)dxdy,$$

$\tau_1(x,y) = x$, $\tau_2(x,y) = y$, and let $N_k(t) = 1_{\{t \ge \tau_k\}}$, $t \in [0,T]$. Then we can easily see that the assumptions (A-1)-(A-3), (M-1),(M-2) are satisfied by taking $d = 0$, $K = 2$, and $\lambda_k(t) = \lambda_k$, $k = 1,2$. Now let $\alpha_1, \alpha_2 > 0$ and

$$\kappa_1(t) = (\frac{\alpha_1}{\lambda_1} - 1)N_2(t-),$$

$$\kappa_2(t) = (\frac{\alpha_2}{\lambda_2} - 1)N_1(t-).$$

Let ρ_t be a solution to the following SDE.

$$d\rho_t = \rho_{t-}(\sum_{k=1}^2 \kappa_k(t)dM_k(t)), \qquad \rho_0 = 1$$

Then ρ_t is a bounded positive martingale. Let Q be probability measure on Ω given by $dQ = \rho_T dP$. Then

$$\tilde{M}_k(t) = N_k(t) - \int_0^t (1 - N_k(s))\tilde{\lambda}_k(s)ds \qquad t \in [0,T], \quad k = 1,2,$$

are Q-$\{\mathcal{F}_t\}_{t\in[0,T]}$ martingale, where

$$\tilde{\lambda}_1(t) = \lambda_1(1 - N_2(t)) + \alpha_1 N_2(t),$$

$$\tilde{\lambda}_2(t) = \lambda_2(1 - N_1(t)) + \alpha_2 N_1(t),$$

Applying Proposition 3.3 we see that

$$\kappa_1^{\{1\}}(t) = \kappa_1(t), \qquad \kappa_2^{\{1\}}(t) = 0, \qquad a.e. t \in [0,T].$$

So we see that $N_2(t) - \int_0^t (1 - N_2(s))\lambda_2 ds$ is $Q^{\{1\}}$ - $\{\mathcal{F}_t\}_{t\in[0,T]}$ martingale. So we see that

$$Q^{\{1\}}(\tau_2 > t|\mathcal{F}_s) = (1 - N_2(s))\exp(-\lambda_2(t-s)), \qquad t,s \in [0,T] \text{ with } s < t.$$

Applying Corollary 3.6, we see that for $s,t \in [0,T]$ with $s < t$,

$$Q(\tau_1 > t|\mathcal{F}_s) = (1 - N_1(s))E^{Q^{\{1\}}}[\exp(-\int_s^t du(1 + \kappa_1^{\{1\}}(u))\lambda_1)|\mathcal{F}_s^{\{1\}}]$$

$$= (1 - N_1(s))\frac{1}{\lambda_1 + \lambda_2 - \alpha_1}$$

$$(\lambda_2 \exp(-\alpha_1(t-s)) + (\lambda_1 - \alpha_1)\exp(-(\lambda_1 + \lambda_2)(t-s))),$$

which is independent of α_2, naturally.

Similarly we have

$$Q(\tau_2 > t)$$

$$= \frac{1}{\lambda_1 + \lambda_2 - \alpha_2}(\lambda_1 \exp(-\alpha_2 t) + (\lambda_2 - \alpha_2)\exp(-(\lambda_1 + \lambda_2)t), \qquad t \in [0,T]$$

Note that

$$Q(\tau_2 > t) = \exp(-\lambda_2 t), \qquad t \in [0,T], \qquad \text{if } \alpha_2 = \lambda_2,$$

and that

$$Q(\tau_2 > t) \to \exp(-(\lambda_1 + \lambda_2)t), \qquad t \in [0,T], \qquad \text{as } \alpha_2 \to \infty.$$

Since

$$E^Q[\exp(-\int_0^t \tilde{\kappa}_1(s)ds)] = E^Q[\exp(-\lambda_1(t \wedge \tau_2) - \alpha_1(t - t \wedge \tau_2))],$$

we see that

$$Q(\tau_1 > t) \neq E^Q[\exp(-\int_0^t \tilde{\kappa}_1(s)ds)]$$

in general, if $\lambda_1 \neq \alpha_1$.

4. Filtering model

In this section we study a filtering model which was introduced by Duffie-Lando [1]. Let $(\Omega, \mathcal{B}, \{\mathcal{B}_t\}_{t\in[0,T]}, P)$ be a complete probability space with filtration satisfying usual hypothesis. Let (B_t, B'_t), $t \in [0,T]$ be a 2-dimensional $\{\mathcal{B}_t\}_{t\in[0,T]}$ Brownian motion. Let $\sigma_i : [0,T] \times \mathbf{R} \to \mathbf{R}$, $i = 0,1$, $b_0 : [0,T] \times \mathbf{R} \to \mathbf{R}$, $b_1 : [0,T] \times \mathbf{R} \times \mathbf{R} \to \mathbf{R}$ be bounded smooth functions, and let $x_0 > 0$, and $y_0 \in \mathbf{R}$. Now let us think of the following S.D.E.

$$dX_t = \sigma_0(t, X_t)dB_t + b_0(t, X_t)dt \qquad X_0 = x_0$$

Let

$$\tau = \inf\{t \in [0,T];\ X_t = 0\}.$$

Now let us think of the following S.D.E.

$$dY_t = \sigma_1(t, Y_t)dB'_t + b_1(t, X_{t\wedge\tau}, Y_t)dt \qquad Y_0 = y_0$$

We assume that there are $c_0 > 0$ such that

$$\sigma_i(t,x) \geq c_0, \qquad t \in [0,T],\ x \in \mathbf{R},\ i = 0,1.$$

Let

$$\mathcal{G}_t = \sigma\{Y_s;\ s \in [0,t]\} \qquad t \in [0,T],$$

and

$$\mathcal{F}_t = \mathcal{G}_t \vee \sigma\{\tau \wedge t\} \qquad t \in [0,T].$$

We will apply the previous results to this situation.

Let

$$F_i(t,x) = \int_0^x \sigma_i(t,y)^{-1}dy, \qquad y \in \mathbf{R},\ i = 0,1.$$

Then $F_i(t,\cdot) : \mathbf{R} \to \mathbf{R}$ $i = 0,1$, are diffeomorphisms.

Let $\tilde{X}_t = F_0(t, X_t)$, $\tilde{Y}_t = F_1(t, Y_t)$, $t \in [0,T]$. Then we have

$$d\tilde{X}_t = dB_t + \beta_0(t, \tilde{X}_t)dt, \qquad \tilde{X}_0 = \tilde{x}_0$$

$$d\tilde{Y}_t = dB'_t + \beta_1(t, \tilde{X}_{t\wedge\tau}, \tilde{Y}_t)dt, \qquad \tilde{Y}_0 = \tilde{y}_0.$$

Here

$$\beta_0(t, F_0(t,x)) = \frac{\partial F_0}{\partial t}(t,x) + \sigma_0(t,x)^{-1}b_0(t,x) + \frac{1}{2}\frac{\partial \sigma_0}{\partial x}(t,x),$$

$$\beta_1(t, F_0(t,x), F_1(t,y)) = \frac{\partial F_1}{\partial t}(t,y) + \sigma_1(t,y)^{-1}b_1(t,x,y) + \frac{1}{2}\frac{\partial \sigma_1}{\partial y}(t,y)$$

and

$$\tilde{x}_0 = F(0,\cdot)^{-1}(x_0), \qquad \tilde{y}_0 = F(0,\cdot)^{-1}(y_0).$$

Then it is obvious that

$$\tau = \inf\{t \in [0,T];\ \tilde{X}_t = 0\}.$$

and

$$\mathcal{G}_t = \sigma\{\tilde{Y}_s;\ s \in [0,t]\} \qquad t \in [0,T].$$

Let

$$\gamma(t,x,y) = \beta_1(t,x,y) - \beta_1(t,0,y), \qquad t \in [0,T],\ x,y \in \mathbf{R},$$

$$\rho = \exp(-\int_0^T (\beta_0(\tilde{X}_t)dB_t + \gamma(t,\tilde{X}_{t\wedge\tau},\tilde{Y}_t)d\tilde{B}_t)$$
$$-\frac{1}{2}\int_0^T (|\beta_0(\tilde{X}_t)|^2 + |\gamma(t,\tilde{X}_{t\wedge\tau},\tilde{Y}_t)|^2)dt),$$

and let $\tilde{P}$ is a probability measure on $(\Omega,\mathcal{B})$ given by $d\tilde{P} = \rho dP$. Then $\tilde{X}$ and $\tilde{Y}$ are independent under $\tilde{P}$ and $\tilde{X}_t$ and $\tilde{B}'_t = \tilde{Y}_t - \int_0^t \beta_1(s,0,\tilde{Y}_s)ds$ are $P - \{\mathcal{B}_t\}_{t\in[0,T]}$ Brownian motions. So $\{\tilde{B}'_t\}_{t\in[0,T]}$ is a Brownian base of $(\Omega,\{\mathcal{G}_t\}_{t\in[0,T]},\tilde{P})$, and τ is independent from $\mathcal{G}_T$ under $\tilde{P}$. Thus we see that under the probability measure $\tilde{P}$ Assumptions (A-1)-(A-3), (M-1) and (M-2) are satisfied by letting $d=1$, $N=1$ and

$$\lambda(t) = -q(t)^{-1}\frac{d}{dt}q(t), \qquad t \in [0,T],$$

where

$$q(t) = \int_t^\infty ds \frac{\tilde{x}_0}{\sqrt{2\pi s^3}}\exp(-\frac{\tilde{x}_0^2}{2s}), \qquad t > 0.$$

Since the measures P and $\tilde{P}$ are equivalent, we can apply the results in Section 3 .

Let

$$g(t,x,y) = \frac{1}{\sqrt{2\pi s}}(\exp(-\frac{(x-y)^2}{2t}) - \exp(-\frac{(x+y)^2}{2t})), \qquad t>0,\ x,y>0,$$

which is the fundamental solution of heat equation with the Dirichlet boundary condition. Then we have

$$q(t) = \int_0^\infty dy\ g(t,\tilde{x}_0,y), \qquad t > 0.$$

Note that

$$\frac{\partial}{\partial x}\log g(t,x,y) = -\frac{x-y}{t} + \frac{1}{x}\varphi(\frac{2xy}{t}), \qquad x,y>0, t>0$$

where $\varphi : \mathbf{R} \to \mathbf{R}$ is a smooth function given by

$$\varphi(x) = \frac{xe^{-x}}{1-e^{-x}}, \quad x \neq 0, \quad \text{and} \quad \varphi(0) = 1.$$

Now let $Z(t) = Z^{s,y}_{0,x}(t)$, $t \in [0,s)$, $x > 0$, $y \geq 0$, $s > 0$, be a solution to the following S.D.E.

$$dZ(t) = dB_t - \frac{Z(t)-y}{s-t}dt + \frac{1}{Z(t)}\varphi(\frac{2Z(t)y}{s-t})dt, \quad t \in [0,s), \qquad Z(0) = x.$$

Then one can see that

$$\inf_{t\in[0,u]} Z^{s,y}_{0,x}(t) > 0, \qquad \text{for all } u \in [0,s),$$

and

$$Z^{s,y}_{0,x}(t) \to y, \qquad t \uparrow s,$$

P-a.s. for any $x > 0$, $y \geq 0$, and $s > 0$. So we may define that $Z^{s,y}_{0,x}(s) = y$.

Let $\nu^{s,y}_{0,x}$, $x > 0$, $y \geq 0$, $s > 0$, be the probability law of $\{Z^{s,y}_{0,x}(t)\}_{t\in[0,s]}$ under P, which is a probability measure on $C([0,s];\mathbf{R})$.

Then we have the following.

Proposition 4.1. *For any $s \in [0,T]$ and a bounded continuous function $f : C([0,s];\mathbf{R}) \to \mathbf{R}$,*

$$E^{\tilde{P}}[f(\{\tilde{X}_{t\wedge\tau}\}_{t\in[0,s]})|\tau \wedge s]$$

$$= 1_{\{\tau>s\}}q(t)^{-1}\int_0^\infty dy\; g(s,\tilde{x}_0,y)\int_{C([0,s];\mathbf{R})} f(w)\nu^{s,y}_{0,\tilde{x}_0}(dw)$$

$$+1_{\{s\geq\tau\}}\int_{C([0,\tau];\mathbf{R})} f(\psi_{\tau,s}(w))\nu^{\tau,0}_{0,\tilde{x}_0}(dw)$$

where $\psi_{t,s}(w)(u) = w(u)$, $u \in [0,t]$, and $\psi_{t,s}(w)(u) = 0$, $u \in (t,s]$.

As a Corollary to this Proposition, we have the following.

Proposition 4.2. *For any $t \in [0,T]$*

$$\rho_t = E^{\tilde{P}}[\frac{dP}{d\tilde{P}}|\mathcal{F}_t]$$

$$= 1_{\{\tau>t\}}q(t)^{-1}\int_0^\infty dy\; g(t,\tilde{x}_0,y)\int_{C([0,t];\mathbf{R})} G(t,w;\tilde{Y})\nu^{s,y}_{0,\tilde{x}_0}(dw)$$

$$+1_{\{t\geq\tau\}}\int_{C([0,\tau];\mathbf{R})} G(t,\psi_{\tau,t}(w);\tilde{Y})\nu^{\tau,0}_{0,\tilde{x}_0}(dw)$$

where

$$G(t,w;\tilde{Y})$$

$$= \exp(\int_0^t (\beta_0(s,w(s))dw(s) + \gamma(s,w(s),\tilde{Y}_s)d\tilde{B}'_s)$$

$$-\frac{1}{2}\int_0^t (|\beta_0(w(s))|^2 + |\gamma(s,w(s),\tilde{Y}_s)|^2)ds),$$

In particular,

$$d\rho_t = \rho_{t-}(h(t)d\tilde{B}'_t + \kappa(t)dM_t),$$

where

$$M_t = N_t - \int_0^t (1 - N_s)\lambda(s)ds, \qquad N_t = 1_{\{t \geq \tau\}},$$

$$h(t)$$

$$= \rho_{t-}^{-1} 1_{\{\tau \geq t\}} q(t)^{-1} \int_0^\infty dy \gamma(t, y, \tilde{Y}_t) g(t, \tilde{x}_0, y) \int_{C([0,t];\mathbf{R})} G(t, w; \tilde{Y}) \nu_{0,\tilde{x}_0}^{s,y}(dw),$$

and

$$\kappa(t) = 1_{\{\tau \geq t\}} (\rho_{t-}^{-1} \int_{C([0,t];\mathbf{R})} G(t, w; \tilde{Y}) \nu_{0,\tilde{x}_0}^{t,0}(dw) - 1)$$

Remark 4.4. By Proposition 3.4, we see that $N_t - \int_0^t (1 + \kappa(s))\lambda(s)ds$ is a martingale under P. By noting that

$$\frac{d}{dt} q(t) = -\lim_{y \downarrow 0} \frac{1}{2y} g(t, \tilde{x}_0, y)$$

we see that

$$(1 + \kappa(t))\lambda(t) = \lim_{x \downarrow 0} \frac{1}{2x} \frac{\tilde{P}(\tilde{X}_t \in dx | \mathcal{F}_t)}{dx} \qquad \text{on } \{\tau > t\},$$

which coincides with a consequence of Duffie-Lando [1] of course. But the equation (1.1) may not hold in our model because of the existence of $\kappa(t)$.

Acknowledgement. The author thanks Nakahiro Yoshida, Kimiaki Aonuma and Koji Inui for usuful discussion and comments.

References

1. Duffie, D., Lando, D.: Term structures of credit spreads with incomplete accounting information. Preprint (1997)
2. Duffie, D., Schroder, M., Skidas, C.: Recursive valuation of defaultable securities and the timing of resolution of uncertainty. Annals of Applied Probability **6**, 1075-1090 (1996)
3. Duffie, D., Singleton, K.: Modelling term structures of defaultable bonds. Working Paper, Graduate School of Business, Stanford University
4. Jarrow, R., Turnbull, S.: Pricing options on financial securities subject to default risk. Journal of Finance **50**, 53-86 (1995)

Adv. Math. Econ. 1, 83–97 (1999)

Evaluation of yield spread for credit risk

Hiroshi Shirakawa*

Department of Industrial Engineering and Management, Tokyo Institute of Technology, 2-12-1 Oh-Okayama, Meguro-ku, Tokyo 152-8552, Japan
(e-mail: sirakawa@me.titech.ac.jp)

Received: April 7, 1998

JEL classification: G120, G130, G320, G330

Mathematics Subject Classification (1991): 60H10, 60H30, 90H09

Summary. We study the rational evaluation of yield spread for defaultable credit with fixed maturity. The default occurs when the asset value hits a given fraction of the nominal credit value. The yield spread is continuously accumulated to the initial credit as an insurance fee for future default. By the rational credit pricing, we prove the unique existence of equilibrium yield spread which satisfies the arbitrage free property. Furthermore we show that this spread yield is independent of the choice of interest rate process. For the quantitative study of rational yield spread, we derive an explicit analytic formula for the equilibrium and show numerical example for various parameters.

Key words: Credit risk, yield spread, balance sheet, equivalent martingale measure, sensitivity analysis.

1. Introduction

Merton [M] is the pioneer who had studied the yield spread for defaultable bond in the systematic way. Merton treated the credit risk as a cap for the maturity face value of credit by the firm's total asset value. In the Merton's original approach, the default occurs only at the maturity of credit. This enables us to evaluate the defaultable bond as a Black-Scholes [B-S] type formula. However in the actual setting, early default occurs even for the fixed maturity credit. Black-Cox [B-C] relaxed the default event so that it occurs when the asset value hits given fixed lower bound. This enables us to derive the reasonable credit spread for the actual corporate debt market. Furthermore Longstaff-Schwartz [L-S] generalized the Black-Cox model so that the default free interest rate process follows Vasicek [V] type model. In

* This research is partly supported by the Industrial Bank of Japan.

their approach, the analytical evaluation of defaultable bond is difficult since the stochastic interest rate affects the arbitrage free price of the yield spread.

In this paper, we consider the default event occurs at the first hitting time for the nominal credit value which includes the accumulated yield spread for the credit risk. For this default model, we can evaluate the arbitrage free yield spread analytically. Furthermore this enables us to separate the yield spread analysis from the default free interest rate process. Therefore once we have derived the arbitrage free yield spread under the constant interest rate model, we can apply the result for any default free interest rate process.

The paper is organized as follows. In Section 2, we consider the arbitrage free defaultable credit evaluation under the given yield spread. In Section 3, we prove the existence of unique yield spread which satisfies the arbitrage free property and derive the explicit credit pricing formula for the given yield spread. Also we prove the nonnegativity of the equilibrium forward spread and exhibit the dependence of the equilibrium yield spread for various risk parameters through the numerical example. Finally in Section 4, we show the independence of the yield spread analysis from the default free interest rate process.

2. Arbitrage free credit risk yield spread

We consider a *complete* security market model with two asset classes, riskless and risky assets. Let $(\Omega, (\mathcal{F}_t)_{0 \leq t}, P)$ be a filtered probability space satisfying the usual condition and let $\boldsymbol{W} = \{W_t \ ; \ 0 \leq t\}$ be a Wiener process under P. We suppose that $\mathcal{F}_t$ is generated by $\{W_u; 0 \leq u \leq t\}$, the riskless interest rate r is constant, and the debt maturity T is fixed. Let V_t be the value of risky asset at time t which follows :

$$\frac{dV_t}{V_t} = \mu dt + \sigma dW_t, \quad \sigma > 0. \tag{2.1}$$

We consider that the credit risk yield spread is continuously accumulated to the nominal credit value as the insurance fee for future default. The default is triggered when the asset value hits a given fraction of the nominal credit value. Let ϕ_T denote the credit risk yield spread for the T-maturity credit. Then the T-maturity nominal credit value $D_{t,T}$ is given by

$$D_{t,T} = D_0 \exp\{(r + \phi_T)t\}, \quad 0 \leq t \leq T. \tag{2.2}$$

Default is triggered when the asset value V_t falls below the $1 - \beta$ nominal credit value, where $\beta \ \in [0.1)$, denotes the uncollected ratio of the nominal credit. The object of this paper is to evaluate the arbitrage free credit risk yield spread ϕ_T^* for the given maturity T.

From the general arbitrage free option pricing theory (e.g., Harrison-Pliska [H-P1, H-P2]), the rational European option price is given as the expected discounted maturity payoff under some equivalent martingale measure. Since we consider only one risky asset, the model becomes complete and the equivalent martingale measure P^* on $(\Omega, \mathcal{F}_T)$ is uniquely defined by the Radon-Nikodym derivative :

$$\rho_T = \left.\frac{dP^*}{dP}\right|_{\mathcal{F}_T} = \exp\left\{-\zeta W_T - \frac{1}{2}\zeta^2 T\right\}, \quad \zeta = \frac{\mu - r}{\sigma}. \tag{2.3}$$

Since $E[\rho_T] = 1$, P^* is actually well defined by (2.3). From the Girsanov's theorem, the drifted Wiener process :

$$W_t^* = W_t + \zeta t,$$

becomes a Wiener process under P^*. Since the asset value V_t is given by :

$$V_t = V_0 \exp\left\{\left(r - \frac{1}{2}\sigma^2\right) t + \sigma W_t^*\right\}, \tag{2.4}$$

its discounted value process $\bar{V}_t = e^{-rt} V_t$ becomes martingale under P^*.

Let α be the ratio of initial credit for initial total asset, i.e. $\alpha = \frac{D_0}{V_0}$. To avoid the trivial default, we assume that the initial state is non-default, i.e. $\alpha \in [0, 1)$. The default time τ is given by the stopping time :

$$\tau = \inf\{t \geq 0 \; ; \; V_t \leq (1 - \beta) D_{t,T}\}. \tag{2.5}$$

From (2.4) through (2.6), τ results in

$$\tau = \inf\{t \; ; \; \sigma W_t^* - \xi_+ t \leq -\gamma\}, \tag{2.6}$$

where

$$\begin{cases} \xi_+ & = & \phi_T + \frac{1}{2}\sigma^2, \\ \gamma & = & -\log[\alpha(1 - \beta)] > 0. \end{cases} \tag{2.7}$$

Here we study the arbitrage free credit side value of balance sheet with the risky asset V_t financed by the nominal credit value $D_{t,T}$. First, we consider the capital value for stockholders using the arbitrage free pricing approach. When default does not occur until the maturity of credit, T-liquidation payoff of the firm is $(V_T - D_{T,T})_+$ for it's stockholders. However they can get no liquidation value if the default occurs since $\beta \geq 0$. Then the arbitrage free price of the initial net asset $C_{0,T}^*$ is given by

$$C_{0,T}^* = E^*[e^{-rT} 1_{\{\tau \geq T\}} (V_T - D_{T,T})_+]. \tag{2.8}$$

By the same way, we can evaluate the arbitrage free T-maturity debt value. If default does not occur until maturity T, the liquidation payoff for the creditors is given by $\min\{V_T, D_{T,T}\}$. If default occurs, the debt is refunded only by the liquidation value $V_\tau = (1 - \beta) D_{\tau,T}$ at the default time τ. Hence the arbitrage free initial credit price $D_{0,T}^*$ is given by

$$D^*_{0,T} = E^*[e^{-rT}1_{\{\tau \geq T\}} \min\{D_{T,T}, V_T\} + e^{-r\tau}1_{\{\tau<T\}}V_\tau]. \quad (2.9)$$

Then the credit side of the arbitrage free balance sheet becomes

$$\begin{aligned} & C^*_{0,T} + D^*_{0,T} \\ = \;& E^*[e^{-rT}1_{\{\tau\geq T\}}(\max\{D_{T,T}, V_T\} - D_{T,T})] \\ & + E^*[e^{-rT}1_{\{\tau\geq T\}} \min\{D_{T,T}, V_T\} + e^{-r\tau}1_{\{\tau<T\}}V_\tau] \\ = \;& E^*[e^{-rT}1_{\{\tau\geq T\}}V_T + e^{-r\tau}1_{\{\tau<T\}}V_\tau] \\ = \;& E^*[e^{-r(\tau\wedge T)}V_{\tau\wedge T}] \\ = \;& V_0. \end{aligned}$$

The last equality follows from the optional sampling theorem [R-Y] and the martingale property of $\bar{V}_t$ under P^*. We can show the following dependence of the arbitrage free debt value $D^*_{0,T}$ for the various parameters.

Lemma 2.1.

(1) $\dfrac{\partial D^*_{0,T}}{\partial \phi_T} > 0$,

(2) $\dfrac{\partial D^*_{0,T}}{\partial \beta} < 0$,

(3) $\dfrac{\partial D^*_{0,T}}{\partial r} = 0$.

Proof. From (2.9), we have

$$\begin{aligned} D^*_{0,T} &= E^*[\min\{D_0 e^{\phi_T T}, V_0 e^{-\frac{1}{2}\sigma^2 T + \sigma W^*_T}\}1_{\{\tau\geq T\}} + D_0(1-\beta)e^{\phi_T \tau}1_{\{\tau<T\}}] \\ &= D_0\left\{e^{\phi_T T}E^*\left[1_{\{\tau\geq T\}}\min\left\{1, \alpha^{-1}e^{\sigma\tilde{W}^*_T}\right\}\right] + (1-\beta)E^*[e^{\phi_T\tau}1_{\{\tau<T\}}]\right\}, \end{aligned}$$

where

$$\tilde{W}^*_t = W^*_t - \left(\frac{\phi_T}{\sigma} + \frac{1}{2}\sigma\right)t.$$

Let $\tau(\phi') = \tau|_{\phi_T=\phi'}$ and $\tau(\beta') = \tau|_{\beta=\beta'}$. Then from the definition of default,

$$\begin{cases} \tau(\phi_1) > \tau(\phi_2) \text{ for } \phi_1 < \phi_2, \\ \tau(\beta_1) < \tau(\beta_2) \text{ for } \beta_1 < \beta_2. \end{cases}$$

(1) Let $0 \leq \phi_1 < \phi_2$, then

$$1_{\{\tau(\phi_1)<T\}}\bar{V}_{\tau(\phi_1)} = 1_{\{\tau(\phi_1)<T\}}D_0 e^{\phi_1\tau(\phi_1)} = 1_{\{\tau(\phi_1)<T\}}\min\{D_0e^{\phi_1 T}, \bar{V}_{\tau(\phi_1)}\}.$$

Therefore the arbitrage free credit value for ϕ_1 is given by

$$\begin{aligned} D^*_{0,T}(\phi_1) &= E^*[1_{\{\tau(\phi_1)\geq T\}}\min\{D_0e^{\phi_1 T}, \bar{V}_T\} + 1_{\{\tau(\phi_1)<T\}}\bar{V}_{\tau(\phi_1)}] \\ &= E^*[\min\{D_0e^{\phi_1 T}, \bar{V}_{\tau(\phi_1)\wedge T}\}]. \end{aligned}$$

Since $\bar{V}_t$ is martingale under P^* and $f(x) = \min\{D_0e^{\phi_1 T}, x\}$ is strict concave, $Y_t = f(\bar{V}_t)$ becomes strict supermartingale under P^*. Then

$$\begin{aligned}
D^*_{0,T}(\phi_1) &= E^*[Y_{\tau(\phi_1)\wedge T}] \\
&= E^*[E^*[Y_{\tau(\phi_1)\wedge T}|\mathcal{F}_{\tau(\phi_2)\wedge T}]] \\
&< E^*[Y_{\tau(\phi_2)\wedge T}] \\
&= E^*[\min\{D_0 e^{\phi_1 T}, \bar{V}_{T\wedge\tau(\phi_2)}\}] \\
&< E^*[\min\{D_0 e^{\phi_2 T}, \bar{V}_{T\wedge\tau(\phi_2)}\}] \\
&= D^*_{0,T}(\phi_2).
\end{aligned}$$

Hence we get $D^*_{0,T}(\phi_1) < D^*_{0,T}(\phi_2)$ for $\phi_1 < \phi_2$.
(2) Let $\beta_1 < \beta_2$. Since Y_t is strict supermartingale under P^*,

$$\begin{aligned}
D^*_{0,T}(\beta_2) &= E^*[\min\{D_0 e^{\phi T}, \bar{V}_{\tau(\beta_2)\wedge T}\}] \\
&= E^*[Y_{\tau(\beta_2)\wedge T}] \\
&= E^*[E^*[Y_{\tau(\beta_2)\wedge T}|\mathcal{F}_{\tau(\beta_1)\wedge T}]] \\
&< E^*[Y_{\tau(\beta_1)\wedge T}] \\
&= E^*[\min\{D_0 e^{\phi T}, \bar{V}_{\tau(\beta_1)\wedge T}\}] \\
&= D^*_{0,T}(\beta_1).
\end{aligned}$$

Hence we have $D^*_{0,T}(\beta_1) > D^*_{0,T}(\beta_2)$ for $\beta_1 < \beta_2$.
(3) Since $\bar{V}_t$ and τ does not depend on r, $D^*_{0,T}$ is independent of r. □

3. Existence of uniqueequilibrium yield spread

Here we consider the rational yield spread pricing which is consistent with the arbitrage free property. As shown in Section 2, the initial balance sheet satisfies the arbitrage free property if and only if the arbitrage free credit value $D^*_{0,T}$ coincides with the initial nominal credit value D_0 for the given yield spread ϕ_T. This leads us to the following equilibrium definition for the rational yield spread.

Definition 3.1. *We call ϕ^*_T is the equilibrium yield spread for T-maturity when $D_0 = D^*_{0,T}|_{\phi_T=\phi^*_T}$ holds.*

After the credit contract, the credit risk yield spread is continuously accumulated to the nominal credit value as the insurance fee for future default. However this spread is equal to the price of credit risk under the equilibrium. Hence the value of the firm does not affected by the existence of the credit contract. To study the equilibrium condition, let us define

$$F(\phi) \triangleq \frac{D^*_{0,T}(\phi)}{D_0}, \tag{3.1}$$

where $D^*_{0,T}(\phi) = D^*_{0,T}|_{\phi_T=\phi}$. The equilibrium yield spread ϕ^*_T is given as the solution of $F(\phi) = 1$. Then we can guarantee the existence and uniqueness of equilibrium yield spread ϕ^*_T as follows.

Theorem 3.1. *Assume that $V_0 \geq D_0$, then there exists a unique strictly positive equilibrium yield spread ϕ_T^* for the T-maturity credit, which is independent of the interest rate r.*

Proof. When $\phi = 0$,

$$\begin{aligned} D_{0,T}^*(0) &= E^*[e^{-rT}1_{\{\tau \geq T\}} \min\{D_{T,T}, V_T\} + e^{-r\tau}1_{\{\tau < T\}} V_\tau] \\ &= E^*[1_{\{\tau \geq T\}} \min\{D_0, \bar{V}_T\} + 1_{\{\tau < T\}} \bar{V}_\tau] \\ &= E^*[1_{\{\tau \geq T\}} \min\{D_0, \bar{V}_T\} + 1_{\{\tau < T\}} \min\{D_0, \bar{V}_T\}] \\ &= E^*[\min\{D_0, V_{\tau \wedge T}\}] \\ &< D_0. \end{aligned}$$

Then we have $F(0) = \frac{D_{0,T}^*(0)}{D_0} < 1$. On the other hand, since $\lim_{\phi \to +\infty} \tau(\phi) = 0$ and $\alpha \leq 1$,

$$\lim_{\phi \to +\infty} D_{0,T}^*(\phi) = V_0 \geq D_0,$$

and hence $\lim_{\phi \to +\infty} F(\phi) \geq 1$. Furthermore $F(\cdot)$ is continuous and strictly increasing, i.e.,

$$\frac{\partial F(\phi)}{\partial \phi} = \frac{1}{D_0} \frac{\partial D_{0,T}^*(\phi)}{\partial \phi} > 0.$$

Then $\exists!\ \phi^*$ such that $F(\phi^*) = 1$. □

In the case of standard Black-Scholes [B-S] type model, we can derive the function $F(\cdot)$ explicitly. Hence we can study the behavior of the equilibrium yield spread numerically.

Lemma 3.1.

$F(\phi)$

$$\begin{aligned} = \alpha^{-1} & \left[\begin{array}{l} \Psi\left(\frac{\gamma - (\phi - \frac{1}{2}\sigma^2)T}{\sigma\sqrt{T}}\right) - \Psi\left(\frac{\eta - (\phi - \frac{1}{2}\sigma^2)T}{\sigma\sqrt{T}}\right) \\ -e^{\frac{2\gamma}{\sigma^2}(\phi - \frac{1}{2}\sigma^2)} \left\{ \Psi\left(\frac{-\gamma - (\phi - \frac{1}{2}\sigma^2)T}{\sigma\sqrt{T}}\right) - \Psi\left(\frac{\eta - 2\gamma - (\phi - \frac{1}{2}\sigma^2)T}{\sigma\sqrt{T}}\right) \right\} \end{array} \right] \\ & + e^{\phi T} \left\{ \Psi\left(\frac{\eta - (\phi + \frac{1}{2}\sigma^2)T}{\sigma\sqrt{T}}\right) - e^{\frac{2\gamma}{\sigma^2}(\phi + \frac{1}{2}\sigma^2)} \Psi\left(\frac{\eta - 2\gamma - (\phi + \frac{1}{2}\sigma^2)T}{\sigma\sqrt{T}}\right) \right\} \\ & + (1-\beta) \left\{ e^{\frac{2\gamma\phi}{\sigma^2}} \Psi\left(\frac{-\gamma - (\phi - \frac{1}{2}\sigma^2)T}{\sigma\sqrt{T}}\right) + e^{\gamma} \Psi\left(\frac{-\gamma + (\phi - \frac{1}{2}\sigma^2)T}{\sigma\sqrt{T}}\right) \right\}, \end{aligned} \tag{3.2}$$

where

$$\begin{cases} \Psi(\cdot) \ : \ \text{the standard normal distribution function}, \\ \alpha \in (0,1], \\ \beta \in [0,1), \\ \gamma = -\log[\alpha(1-\beta)] > \eta, \\ \eta = -\log \alpha > 0. \end{cases}$$

Proof. From (2.9) and (3.1),

$$F(\phi) = e^{\phi T} E^*[1_{\{\tau \geq T\}} \min\{1, \alpha^{-1} e^{\sigma \tilde{W}_T^*}\}] + (1-\beta) E^*[e^{\phi \tau} 1_{\{\tau \leq T\}}]. \quad (3.3)$$

Now let $f(x,y)$ be the joint density function for $\left(\min_{0 \leq \tau \leq T} \sigma \tilde{W}_t^*, \sigma \tilde{W}_T^*\right)$. Then $f(x,y)$ is given by

$$\begin{aligned} f(x,y) &= \frac{\partial^2}{\partial x \partial y} P^* \left[\min_{0 \leq t \leq T} \sigma \tilde{W}_e^* \leq x, \sigma \tilde{W}_T^* \leq y \right] \\ &= -\frac{2(2x-y)}{\sqrt{2\pi (\sigma^2 T)^3}} e^{-\frac{(2x-y-\xi_+ T)^2}{2\sigma^2 T} - \frac{2x\xi_+}{\sigma^2}}, \quad x \leq y, \ x \leq 0. \end{aligned}$$

Also let $\alpha^{-1} e^{-y^*} = 1$, i.e. $y^* = -\log \alpha \geq 0$ and $\xi_- = \phi - \frac{1}{2}\sigma^2$. Then

$$\begin{aligned} & E^*[1_{\{\tau \geq T\}} \min\{1, \alpha^{-1} e^{\sigma \tilde{W}_T^*}\}] \\ &= E^* \left[1_{\{\min_{0 \leq t \leq T} \sigma \tilde{W}_t^* \geq -\gamma\}} \min\{1, \alpha^{-1} e^{\sigma \tilde{W}_T^*}\} \right] \\ &= \int_{-\gamma}^{0} \int_{x}^{\infty} \min\{1, \alpha^{-1} e^y\} f(x,y) dy dx \\ &= \int_{-\gamma}^{\infty} \int_{-\gamma}^{y \wedge 0} \min\{1, \alpha^{-1} e^y\} f(x,y) dx dy \\ &= \int_{-\gamma}^{0} \int_{-\gamma}^{y} \min\{1, \alpha^{-1} e^y\} f(x,y) dx dy + \int_{0}^{\infty} \int_{-\gamma}^{0} f(x,y) dx dy \\ &= \int_{-\gamma}^{-y^*} \int_{-\gamma}^{y} \alpha^{-1} e^y f(x,y) dx dy \\ &\quad + \int_{-y^*}^{0} \int_{-\gamma}^{y} f(x,y) dx dy + \int_{0}^{\infty} \int_{-\gamma}^{0} f(x,y) dx dy. \end{aligned} \quad (3.4)$$

The first term of (3.4) is given by

$$\begin{aligned} & \int_{-\gamma}^{-y^*} \int_{-\gamma}^{y} \alpha^{-1} e^y f(x,y) dx dy \\ &= \int_{-\gamma}^{-y^*} \int_{-\gamma}^{y} \alpha^{-1} e^y \frac{-2(2x-y)}{\sqrt{2\pi(\sigma^2 T)^3}} e^{-\frac{(2x-y-\xi_+ T)^2}{2\sigma^2 T} - \frac{2x\xi_+}{\sigma}} dx dy \\ &= \int_{-\gamma}^{-y^*} \frac{1}{\alpha \sqrt{2\pi\sigma^2 T}} \int_{-\gamma}^{y} \frac{\partial}{\partial x} e^{-\frac{1}{2\sigma^2 T}\left((2x-y)^2 + 2Ty\xi_- + \xi_+^2 T^2\right)} dx dy \end{aligned}$$

$$
\begin{aligned}
&= \alpha^{-1}e^{-\phi T}\left\{\begin{array}{l}\int_{-\gamma}^{-y^*}\frac{1}{\sqrt{2\pi\sigma^2T}}e^{-\frac{1}{2\sigma^2T}(y+\xi_-T)^2}dy\\ -e^{\frac{2\gamma\xi_-}{\sigma^2}}\int_{-\gamma}^{-y^*}\frac{1}{\sqrt{2\pi\sigma^2T}}e^{-\frac{1}{2\sigma^2T}(y+2\gamma+\xi_-T)^2}dy\end{array}\right\}\\
&= \alpha^{-1}e^{-\phi T}\left[\begin{array}{l}\Psi\left(\frac{-y^*+\xi_-T}{\sigma\sqrt{T}}\right)-\Psi\left(\frac{-\gamma+\xi_-T}{\sigma\sqrt{T}}\right)\\ -e^{\frac{2\gamma\xi_-}{\sigma^2}}\left\{\Psi\left(\frac{-y^*+2\gamma+\xi_-T}{\sigma\sqrt{T}}\right)-\Psi\left(\frac{\gamma+\xi_-T}{\sigma\sqrt{T}}\right)\right\}\end{array}\right]\\
&= \alpha^{-1}e^{-\phi T}\left[\begin{array}{l}\Psi\left(\frac{\gamma-\xi_-T}{\sigma\sqrt{T}}\right)-\Psi\left(\frac{y^*-\xi_-T}{\sigma\sqrt{T}}\right)\\ -e^{\frac{2\gamma\xi_-}{\sigma^2}}\left\{\Psi\left(\frac{-\gamma-\xi_-T}{\sigma\sqrt{T}}\right)-\Psi\left(\frac{y^*-2\gamma-\xi_-T}{\sigma\sqrt{T}}\right)\right\}\end{array}\right]. \qquad (3.5)
\end{aligned}
$$

The second term of (3.4) is given by

$$
\begin{aligned}
&\int_{-y^*}^{0}\int_{-\gamma}^{y} f(x,y)\,dxdy\\
&= \int_{-y^*}^{0}\int_{-\gamma}^{y} -\frac{2(2x-y)}{\sqrt{2\pi(\sigma^2T)^3}}e^{-\frac{(2x-y-\xi_+T)^2}{2\sigma^2T}-\frac{2x\xi_+}{\sigma^2}}dxdy\\
&= \int_{-y^*}^{0}\frac{1}{\sqrt{2\pi\sigma^2T}}\int_{-\gamma}^{y}\frac{\partial}{\partial x}e^{\frac{1}{\sigma^2}\left(-\frac{1}{2T}(2x-y)^2-y\xi_+-\frac{\xi_+^2}{2}T\right)}dxdy\\
&= \int_{-y^*}^{0}\frac{1}{\sqrt{2\pi\sigma^2T}}\left\{e^{\frac{1}{\sigma^2}\left(-\frac{1}{2T}y^2-y\xi_+-\frac{\xi_+^2}{2}T\right)}-e^{\frac{1}{\sigma^2}\left(-\frac{1}{2T}(y+2\gamma)^2-y\xi_+-\frac{\xi_+^2}{2}T\right)}\right\}dy\\
&= \int_{-y^*}^{0}\frac{1}{\sqrt{2\pi\sigma^2T}}e^{-\frac{1}{2\sigma^2T}(y+\xi_+T)^2}dy\\
&\quad -\int_{-y^*}^{0}\frac{1}{\sqrt{2\pi\sigma^2T}}e^{-\frac{1}{2\sigma^2T}(y+2\gamma+\xi_+T)^2+\frac{2\gamma\xi_+}{\sigma^2}}dy\\
&= \Psi\left(\frac{\xi_+T}{\sigma\sqrt{T}}\right)-\Psi\left(\frac{-y^*+\xi_+T}{\sigma\sqrt{T}}\right) \qquad (3.6)\\
&\quad -e^{\frac{2\gamma\xi_+}{\sigma^2}}\left\{\Psi\left(\frac{2\gamma+\xi_+T}{\sigma\sqrt{T}}\right)-\Psi\left(\frac{-y^*+2\gamma+\xi_+T}{\sigma\sqrt{T}}\right)\right\}.
\end{aligned}
$$

The last term of (3.4) is given by

$$
\begin{aligned}
&\int_{0}^{\infty}\int_{-\gamma}^{0} f(x,y)\,dxdy\\
&= \int_{0}^{\infty}\frac{1}{\sqrt{2\pi\sigma^2T}}\int_{-\gamma}^{0}\frac{\partial}{\partial x}e^{\frac{1}{\sigma^2}\left(-\frac{1}{2T}(2x-y)^2-y\xi_+-\frac{\xi_+^2}{2}T\right)}dxdy\\
&= \int_{0}^{\infty}\frac{1}{\sqrt{2\pi\sigma^2T}}\left\{e^{-\frac{1}{2\sigma^2T}(y+\xi_+T)^2}-e^{-\frac{1}{2\sigma^2T}(y+2\gamma+\xi_+T)^2+\frac{2\gamma\xi_+}{\sigma^2}}\right\}dy\\
&= 1-\Psi\left(\frac{\xi_+T}{\sigma\sqrt{T}}\right)-e^{\frac{2\gamma\xi_+}{\sigma^2}}\left\{1-\Psi\left(\frac{2\gamma+\xi_+T}{\sigma\sqrt{T}}\right)\right\}. \qquad (3.7)
\end{aligned}
$$

Hence from (3.5) through (3.7),

$$\begin{aligned}
& e^{\phi T} \times (3.4) \\
= \;& e^{\phi T}\{(3.5)+(3.6)+(3.7)\} \\
= \;& \alpha^{-1}\left[\begin{array}{l} \Psi\left(\frac{\gamma-\xi_- T}{\sigma\sqrt{T}}\right) - \Psi\left(\frac{y^*-\xi_- T}{\sigma\sqrt{T}}\right) \\ -e^{\frac{2\gamma\xi_-}{\sigma^2}}\left\{\Psi\left(\frac{-\gamma-\xi_- T}{\sigma\sqrt{T}}\right) - \Psi\left(\frac{y^*-2\gamma-\xi_- T}{\sigma\sqrt{T}}\right)\right\} \end{array}\right] \\
& +e^{\phi T}\left\{\Psi\left(\frac{y^*-\xi_+ T}{\sigma\sqrt{T}}\right) - e^{\frac{2\gamma\xi_+}{\sigma^2}}\Psi\left(\frac{y^*-2\gamma-\xi_+ T}{\sigma\sqrt{T}}\right)\right\}. \qquad (3.8)
\end{aligned}$$

Next we shall consider the second term of (3.3). Let $f(x)$ be the density function for τ, that is :

$$f(x) = \frac{\partial}{\partial t}P^*[\tau \le T] = \frac{\gamma}{\sqrt{2\pi\sigma^2 t^3}} e^{-\frac{(-\gamma+\xi_+ t)^2}{2\sigma^2 t}}.$$

Then

$$\begin{aligned}
& E^*[1_{\{\tau\le T\}}e^{\phi\tau}] \\
= \;& \int_0^T e^{\phi t}\frac{\gamma}{\sqrt{2\pi\sigma^2 t^3}} e^{-\frac{(\gamma-\xi_+ t)^2}{2\sigma^2 t}}\,dt \\
= \;& \int_0^T \frac{1}{\sqrt{2\pi}}\left(\frac{\gamma-\xi_- t}{2\sigma\sqrt{t^3}} + \frac{\gamma+\xi_- t}{2\sigma\sqrt{t^3}}\right) e^{-\frac{1}{2\sigma^2 t}(\gamma^2+\xi_-^2 t^2)+\frac{\gamma\xi_+}{\sigma^2}} \\
= \;& -\int_0^T \frac{1}{\sqrt{2\pi}}\frac{\partial}{\partial t}\left(\frac{\gamma+\xi_- t}{\sigma\sqrt{t}}\right) e^{-\frac{1}{2\sigma^2 t}(\gamma+\xi_- t)^2+\frac{2\gamma\phi}{\sigma^2}}\,dt \\
& -\int_0^T \frac{1}{\sqrt{2\pi}}\frac{\partial}{\partial t}\left(\frac{\gamma-\xi_- t}{\sigma\sqrt{t}}\right) e^{-\frac{1}{2\sigma^2 t}(\gamma-\xi_- t)^2+\gamma}\,dt \\
= \;& -e^{\frac{2\gamma\phi}{\sigma^2}}\int_{+\infty}^{\frac{\gamma+\xi_- T}{\sigma\sqrt{T}}} \frac{1}{\sqrt{2\pi}}e^{-\frac{1}{2}x^2}\,dx - e^{\gamma}\int_{+\infty}^{\frac{\gamma-\xi_- T}{\sigma\sqrt{T}}} \frac{1}{\sqrt{2\pi}}e^{-\frac{1}{2}x^2}\,dx \\
= \;& e^{\frac{2\gamma\phi}{\sigma^2}}\left\{1-\psi\left(\frac{\gamma+\xi_- T}{\sigma\sqrt{T}}\right)\right\} + e^{\gamma}\left\{1-\psi\left(\frac{\gamma-\xi_- T}{\sigma\sqrt{T}}\right)\right\} \\
= \;& e^{\frac{2\gamma\phi}{\sigma^2}}\Psi\left(\frac{-\gamma-\xi_- T}{\sigma\sqrt{T}}\right) + e^{\gamma}\Psi\left(\frac{-\gamma+\xi_- T}{\sigma\sqrt{T}}\right). \qquad (3.9)
\end{aligned}$$

From (3.8) and (3.9), we get the desired result (3.2). □

Since the yield spread is caused by the credit risk, it is reasonable to expect that the yield spread increases as the degree of credit risk increases. Hereafter we shall consider the qualitative dependence of the equilibrium yield spread for the various risk parameters.

Theorem 3.2.

(1) $\frac{\partial\phi_T^*}{\partial\beta} > 0$,

(2) $\frac{\partial\phi_T^*}{\partial r} = 0$.

Proof. Since the proof for (1) and (2) are similar, we prove only (1). By the definition,

$$\frac{\partial D^*_{0,T}}{\partial \phi^*_T} d\phi^*_T + \frac{\partial D^*_{0,T}}{\partial \beta} d\beta = dD_0 = 0.$$

Then from Lemma 2.1,

$$\frac{\partial \phi^*_T}{\partial \beta} = -\frac{\frac{\partial D^*_{0,T}}{\partial \beta}}{\frac{\partial D^*_{0,T}}{\partial \phi^*}} > 0. \qquad \square$$

From the definition of α, it is reasonable to expect that $\frac{\partial \phi^*_T}{\partial \alpha} > 0$ holds. However it seems quite difficult to execute the sensitivity analysis for α. Next we consider the increasing property of the total yield spread through the forward spread.

Definition 3.2. *Let us define* θ^*_T *by*

$$\theta^*_T = \frac{\partial}{\partial T}(\phi^*_T T) = \frac{\partial \phi^*_T}{\partial T} T + \phi^*_T.$$

Then we call θ^*_T *the* T*-maturity equilibrium forward spread.*

Then we can show the nonnegativity of the equilibrium forward spread.

Theorem 3.3. *The equilibrium forward spread* θ^*_T *is nonnegative for any maturity* T.

Proof. Let $0 \le T_1 \le T_2$. Then $\tau \wedge T_1 \le \tau \wedge T_2$. Since $\bar{V}_t$ is martingale under P^* and $f(x) = \min\{D_0 e^{\phi^*_{T_2} T_2}, x\}$ is concave, $Y_t = f(\bar{V}_t)$ becomes supermartingale under P^*. Hence

$$\begin{aligned} D_0 &= D^*_{0,T_2} \\ &= E^*[\min\{D_0 e^{\phi^*_{T_2} T_2}, \bar{V}_{\tau \wedge T_2}\}] \\ &= E^*[E^*[Y_{\tau \wedge T_2} | \mathcal{F}_{\tau \wedge T_1}]] \\ &\le E^*[Y_{\tau \wedge T_1}] \\ &= E^*[\min\{D_0 e^{\phi^*_{T_2} T_2}, \bar{V}_{\tau \wedge T_1}\}]. \end{aligned} \tag{3.10}$$

On the other hand,

$$D_0 = D^*_{0,T_1} = E^*[\min\{D_0 e^{\phi^*_{T_1} T_1}, \bar{V}_{\tau \wedge T_1}\}]. \tag{3.11}$$

Hence from (3.10) and (3.11), we have

$$E^*[\min\{D_0 e^{\phi^*_{T_1} T_1}, \bar{V}_{\tau \wedge T_1}\}] \le E^*[\min\{D_0 e^{\phi^*_{T_2} T_2}, \bar{V}_{\tau \wedge T_1}\}].$$

Then we get $\phi^*_{T_1} T_1 \le \phi^*_{T_2} T_2$ for $T_1 \le T_2$ and

$$\theta^*_T = \frac{\partial}{\partial T}(\phi^*_T T) \ge 0. \qquad \square$$

Example 3.1. We show the numerical results of the equilibrium yield spread for various parameters. Let the initial parameters be :

$$\begin{cases} \alpha & = & 0.5, \\ \beta & = & 0.5, \\ T & = & 1, \\ \sigma & = & 0.4. \end{cases}$$

Then we compute the equilibrium yield spread ϕ_T^* when the parameter changes within a given range. Figure 1 through 4 show the equilibrium yield spread for the ratio of credit to the total asset : $0 \leq \alpha < 1$, the default ratio of credit : $0 \leq \beta \leq 1$, the maturity of debt : $0 \leq T \leq 10$ and the volatility of risky asset : $0 < \sigma \leq 1$. As observed in each figures, the equilibrium yield spread increases as each corresponding parameters increases. These dependence is reasonable since each parameters reflect the degree of risk for the default.

4. Generalization of interest rate process

In the previous sections, we assumed the standard Black-Scholes [B-S] type model. Here we generalize the interest rate process $\{r_t \; ; \; 0 \leq t\}$ to be $\{\mathcal{F}_t \; ; \; 0 \leq t\}$-adapted, where $\mathcal{F}_t$ is generated by the 2-dimensional Wiener process $\boldsymbol{W} = \{(W_{1,t}, W_{2,t}) \; ; \; 0 \leq t\}$. As to the risky asset, we assume (2.1) as before except that $W_t := W_{1,t}$. Define the equivalent measure P^* on $(\Omega, \mathcal{F}_T)$ by

$$\rho_T = \left.\frac{dP^*}{dP}\right|_{\mathcal{F}_T} = \exp\left\{ - \sum_{1 \leq i \leq 2} \int_0^T \zeta_{i,t} dW_{i,t} - \frac{1}{2} \sum_{1 \leq i \leq 2} \int_0^T \zeta_{i,t}^2 dt \right\}, \quad (4.1)$$

where

$$\begin{cases} \zeta_{1,t} = \frac{\mu - r_t}{\sigma}, \\ \zeta_{2,t} = \text{arbitrary}. \end{cases}$$

We assume that P^* is well defined by (4.1), i.e. $E[\rho_T] = 1$. Then the drifted 2-dimensional Wiener process :

$$W_{i,t}^* = W_{i,t} + \int_0^t \zeta_{i,u} du.$$

becomes a 2-dimensional Wiener process under P^* by the Girsanov's theorem. Now let $\bar{V}_t$ be the discounted asset value and $\bar{D}_{t,T}$ be the discounted T-maturity credit value, i.e.

$$\begin{cases} \bar{V}_t & = & \exp\{-\int_0^t r_u du\} V_t, \\ \bar{D}_{t,T} & = & \exp\{-\int_0^t r_u du\} D_{t,T}. \end{cases}$$

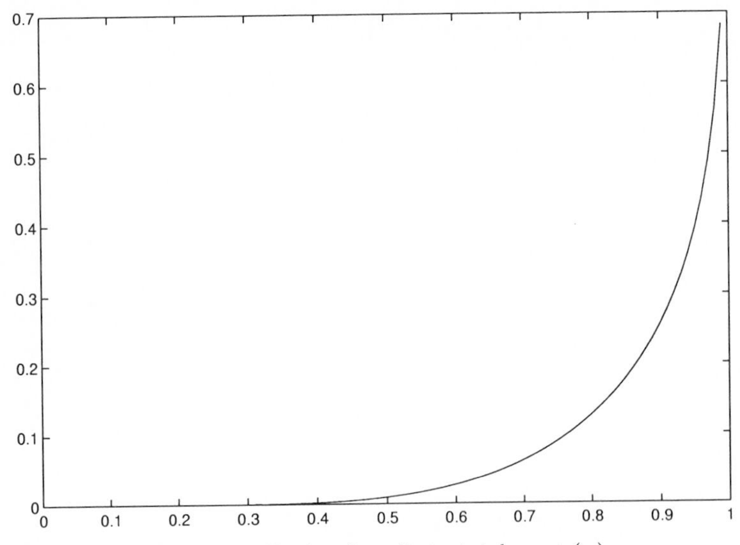

Figure 1. Ratio of credit to total asset (α)

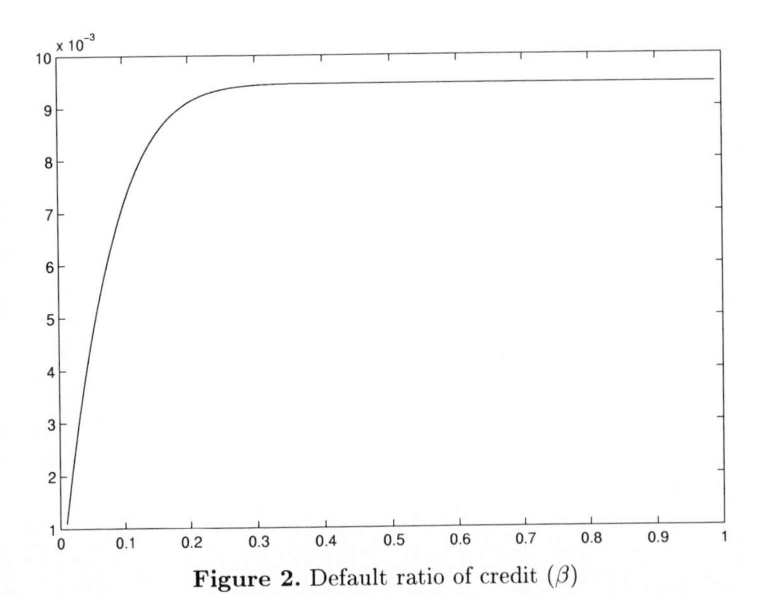

Figure 2. Default ratio of credit (β)

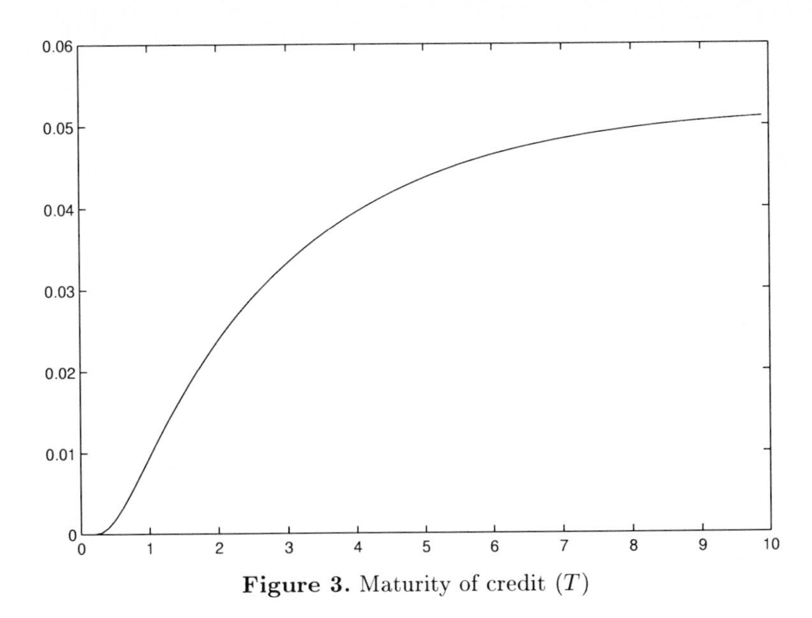

Figure 3. Maturity of credit (T)

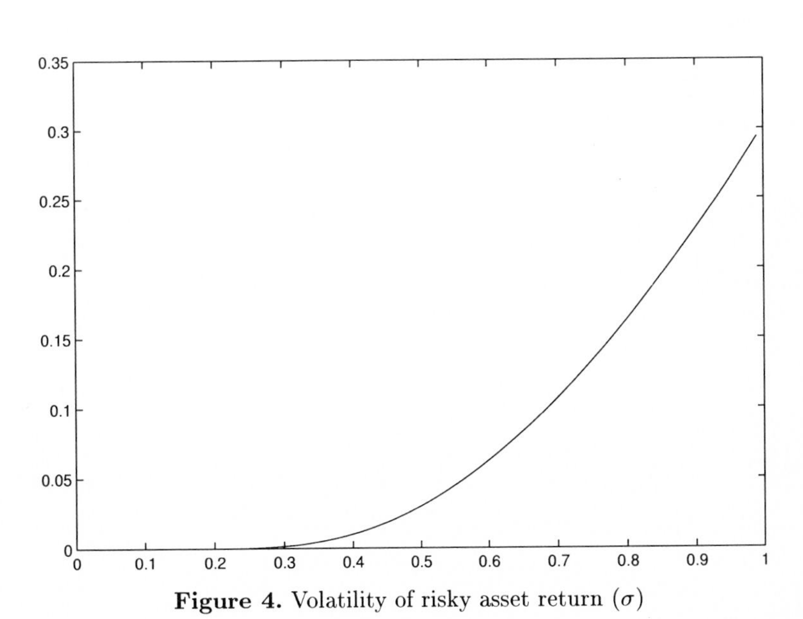

Figure 4. Volatility of risky asset return (σ)

Then $\bar{V}_t$ and $\bar{D}_{t,T}$ are given by

$$\begin{cases} \bar{V}_t &= V_0 \exp\{-\frac{1}{2}\sigma^2 t + \sigma W^*_{1,t}\}, \\ \bar{D}_{t,T} &= D_0 \exp\{\phi_T t\}, \end{cases} \tag{4.2}$$

and hence the default time τ becomes

$$\tau = \inf\{t \; ; \; V_t \leq (1-\beta)D_{t,T}\} = \inf\{t \; ; \; \bar{V}_t \leq (1-\beta)\bar{D}_{t,T}\}. \tag{4.3}$$

This means that τ is independent of the choice of the interest rate process under P^*. Now we shall consider the effect for the equilibrium yield spread from the generalization of model. By the arbitrage free pricing theory [H-P1, H-P2], rational initial credit value $D^*_{0,T}$ is given by

$$D^*_{0,T} = E^*[1_{\{\tau \geq T\}} \min\{\bar{D}_{T,T}, \bar{V}_T\} + 1_{\{\tau < T\}} \bar{V}_\tau]. \tag{4.4}$$

Then from (4.2) through (4.4), $D^*_{0,T}$ does not depend on the interest rate process. Thus we arrive at the following result.

Theorem 4.1. *The equilibrium yield spread ϕ^*_T is independent of the choice of interest rate process as much as the risky asset return process is invariable.* □

Theorem 4.1 guarantees that the risk premium analysis in Sections 1 through 3 are still valid under the general interest rate process. Furthermore, we can show these qualitative analysis are still valid under the more general setting not only for the interest rates, but also for the risky asset investment value process.

References

[B-C] Black, F., Cox, J.C.: Valuing corporate securities - Some effects of bond indenture provisions. J. Finance **31**, 351-367 (1976)
[B-S] Black, F., Scholes, M.: The pricing of options and corporate liabilities. J. Political Economy **81**, 637-654 (1973)
[C-M] Cooper, I., Mello, A.: The default risk of swaps. J. Finance **46**, 597-620 (1991)
[C-R] Cox, J.C., Ross, S.: The valuation of options for alternative stochastic processes. J. Financial Economics **3**, 145-166 (1975)
[H-K] Harrison, J.M., Kreps, D.M.: Martingales and arbitrage in multiperiod security markets. J. Economic Theory **20**, 381-408 (1979)
[H-P1] Harrison, J.M., Pliska, S.R.: Martingales and stochastic integrals in the theory of continuous trading. Stochastic Processes and Their Applications **11**, 381-408 (1981)
[H-P2] Harrison, J.M., Pliska, S.R.: A stochastic calculus model of continuous time trading - Complete markets. Stochastic Processes and Their Applications **13**, 313-316 (1983)

[K-S] Karatzas, I., Shreve, S.E.: Brownian Motion and Stochastic Calculus. Springer-Verlag 1988

[L-S] Longstaff, F.A., Schwartz, E.S.: A simple approach to valuing risky fixed and floating rate debt. J. Finance **50**, 789-819 (1995)

[M] Merton, R.C.: On the pricing of corporate debt - The risk structure of interest rates. J. Finance **29**, 449-470 (1974)

[R-Y] Revuz, D., Yor, M.: Continuous Martingales and Brownian Motion. Springer-Verlag 1991

[V] Vasicek, O.: An equilibrium characterization of the term structure. J. Financial Economics **5**, 177-188 (1977)

Adv. Math. Econ. 1, 99–113 (1999)

Analysis of the asymptotic distance between oscillating functions and their weak limit in L^2

Michel Valadier

Département de Mathématiques — case 051, Université Montpellier II, place Eugène Bataillon, 34 095 Montpellier Cedex 5, France
(e-mail: valadier@math.univ-montp2.fr)

Received: March 2, 1998

JEL classification: C69

Mathematics Subject Classification (1991): 35B40, 28A20, 49J45

Summary. Young measures theory is applied to the understanding of weak convergence without strong convergence in L^2 spaces. The two-scale Young measures permit also to analyse, when it happens, a "modulated periodical" behavior and, in the general case, to get a kind of orthogonal decomposition. Some new examples are discussed.

Key words: Weak convergence, oscillations, asymptotic behavior, Young measures, two-scale convergence

1. Introduction

This paper is devoted to the application of Young measures theory to the understanding of weak convergence in L^2 spaces when strong convergence does not hold (some results would be valid in L^p spaces for $p \in [1, +\infty[$). In any infinite dimensional Hilbert space H there are weakly convergent sequences which do not strongly converge (think of orthonormal sequences). When $H = L^2(\Omega)$ it is well known that such a sequence usually (see the uniform integrability hypothesis in Part 2 of Theorem 1 and the Example after it for the necessity of this hypothesis) "oscillates" on a non negligible part of Ω, that is does not converge almost everywhere.

One line of research looked for sufficient conditions ensuring strong convergence: [Vi], [Bd2], [Ca], [Rz], [Va1], [ACV], [BGJ], [SV1]. When oscillations do occur, except the rather trivial — but basic in homogenization theory (see [Su]) — periodical case, the analysis of the phenomenon is not easy. Young measures, whose introduction goes back to 1937 [Y], were used by L. Tartar [Ta1–2] to obtain limits of non linear functionals and to get an object retaining something of the asymptotic behavior of the oscillating sequence. More recently [Ng], [Al1–2], [E] introduced the two-scale convergence in order to treat new applications in homogenization. This method permits also to analyse a "modulated periodical" behavior. Two-scale Young measures have been defined in [E] and will be used here as well as the new notion of admissible functions introduced in [Va5]. The present paper improves different quantitative results disseminated in previous works of the author: [Va2–3], [SV2] (where the biting lemma is used when uniform integrability is lacking), [Va5] and gives some new examples (specialy the case of i.i.d. random variables, see Proposition 2).

2. Application of classical Young measures

Let Ω be an open subset of $\mathbb{R}^N$ of finite Lebesgue measure $|\Omega|$. The Lebesgue measure on Ω, which is also the N-dimensional Hausdorff measure $\mathcal{H}^N \lfloor_\Omega$, will be denoted by dx. By a *Dirac mass* (or Dirac measure) we mean a measure δ_a concentrating the mass 1 on the point a. Note that if $(\nu_x)_{x\in\Omega}$ is a measurable family of probabilities on $\mathbb{R}$, the set $W = \{x \in \Omega : \nu_x$ is not a Dirac mass$\}$ is measurable because

$$\begin{aligned} W = \{x : \exists (q,q',n) \in \mathbb{Q}^2 \times \mathbb{N}^*, \text{ such that } B(q,1/n) \cap B(q',1/n) \neq \emptyset, \\ \text{satisfying } \nu_x(B(q,1/n)) > 0 \text{ and } \nu_x(B(q',1/n)) > 0\} \\ = \bigcup_{\substack{(q,q',n)\in\mathbb{Q}^2\times\mathbb{N}^* \\ B(q,1/n)\cap B(q',1/n)\neq\emptyset}} \Big[\{x : \nu_x(B(q,1/n)) > 0\} \cap \{x : \nu_x(B(q',1/n)) > 0\}\Big] . \end{aligned}$$

By $\mathrm{frac}(z)$ we denote the fractional part of $z \in \mathbb{R}^N$ (it operates coordinate by coordinate: $\mathrm{frac}(z_1, \ldots, z_N) = (\mathrm{frac}(z_1), \ldots, \mathrm{frac}(z_N))$). Then $\mathrm{frac}(z)$ belongs to $Y := [0,1[^N$.

Definition. *The* Young measure associated to $u_n \in L^2(\Omega)$ *is the measure* ν^n *on* $\Omega \times \mathbb{R}$ *which is the image of* dx *by* $x \mapsto \big(x, u_n(x)\big)$.

A general *Young measure* is a positive measure ν on $\Omega \times \mathbb{R}$ such that $\forall A \in \mathcal{B}(\Omega)$, $\nu(A \times \mathbb{R}) = |A|$ (this means that the projection of ν on Ω is dx). Thanks to a denseness result (which holds because dx is atomless) such a ν is always the limit in the narrow topology defined below of some sequence of Young measures associated to functions. The set of all Young measures on $\Omega \times \mathbb{R}$ will be denoted by $\mathcal{Y}(\Omega, dx; \mathbb{R})$. The *disintegration* of ν is a measurable family $(\nu_x)_{x\in\Omega}$ of probabilities on $\mathbb{R}$ such that, for any function $\psi : \Omega \times \mathbb{R} \to \overline{\mathbb{R}}$ which is positive measurable or ν-integrable,

$$\iint_{\Omega\times\mathbb{R}} \psi \, d\nu = \int_\Omega \Big[\int_\mathbb{R} \psi(x,\lambda)\, d\nu_x(\lambda)\Big]\, dx .$$

One can write

$$\nu = \int_\Omega \delta_x \otimes \nu_x \, dx$$

which is a weak integral of measures whose meaning is as well

$$\forall B \in \mathcal{B}(\Omega \times \mathbb{R}), \quad \nu(B) = \int_\Omega [\delta_x \otimes \nu_x](B) \, dx$$

as

$$\forall \varphi \in \mathcal{C}_c(\Omega \times \mathbb{R}), \quad \iint_{\Omega\times\mathbb{R}} \varphi \, d\nu = \int_\Omega [\int_\mathbb{R} \varphi(x,\lambda) \, d\nu_x(\lambda)] \, dx \, .$$

The disintegration of ν^n (associated to u_n) is $\nu_x^n = \delta_{u_n(x)}$.

The *narrow topology* on $\mathcal{Y}(\Omega, dx; \mathbb{R})$ is the weakest one making continuous the maps $\nu \mapsto \int_{\Omega\times\mathbb{R}} \psi \, d\nu$ where ψ runs through $\mathcal{C}^b(\Omega\times\mathbb{R})$. The same topology can be defined by smaller or larger classes of test functions. In the "smaller direction" the narrow topology coincides with the weak topology defined by test functions belonging to $\mathcal{C}_c(\Omega) \otimes \mathcal{C}_c(\mathbb{R})$ (see the proof in [Va3, p.362]). For the "larger direction" we first define a *Carathéodory integrand* as a real function on $\Omega \times \mathbb{R}$ which is separately measurable on Ω and continuous on $\mathbb{R}$. The set of all bounded Carathéodory integrand is denoted by $\mathcal{C}th^b(\Omega;\mathbb{R})$. The narrow topology on $\mathcal{Y}(\Omega, dx; \mathbb{R})$ coincide with the weak topology defined by test functions belonging to[1] $\mathcal{C}th^b(\Omega;\mathbb{R})$ (references: see [Bd3], [Va3, Th.3]). We will use in the sequel the fact that unbounded Carathéodory integrands can be used as soon as some uniform integrability condition is satisfied.

Definition. *One says that u_n converges* purely weakly *to u_∞ on $W \in \mathcal{B}(\Omega)$ if the restrictions $u_{n|W}$ converge weakly (that is $\sigma(L^2, L^2)$) to $u_{\infty|W}$ and, for any non negligible measurable subset A of W, not any subsequence of $(u_{n|A})_n$ converges in measure to $u_{\infty|A}$.*

In the following $\mathrm{bar}(\nu_x)$ denotes the barycenter of the first order probability ν_x on $\mathbb{R}$: $\mathrm{bar}(\nu_x) = \int_\mathbb{R} \lambda \, d\nu_x(\lambda)$.

Theorem 1. Let $(u_n)_n$ be a bounded sequence in $L^2_\mathbb{R}(\Omega, dx)$. There exist a subsequence $(u'_n)_n$, a function $u_\infty \in L^2_\mathbb{R}(\Omega, dx)$ and a Young measure ν on $\Omega \times \mathbb{R}$ with the following properties. Let $(\nu_x)_{x\in\Omega}$ denote the disintegration of ν, $M = \{x \in \Omega : \nu_x \text{ is a Dirac mass}\}$ and $W := M^c$. Then:

1a) $(\nu'_n)_n$ converges to ν that is, for any $\psi \in \mathcal{C}th^b(\Omega;\mathbb{R})$,

$$\int_\Omega \psi(x, u'_n(x)) \, \mu(dx) \longrightarrow \iint_{\Omega\times\mathbb{R}} \psi \, d\nu \, .$$

Moreover, for almost every x, ν_x is of first order and $u_\infty(x) = \mathrm{bar}(\nu_x)$. In particular $\nu_x \overset{\text{a.e.}}{=} \delta_{u_\infty(x)}$ on M;

1b) u'_n converges weakly (i.e. $\sigma(L^2, L^2)$) to u_∞;

[1] In the case of an abstract measured space $(\Omega, \mathcal{F}, \mu)$ this is the good way to define the "narrow topology" on $\mathcal{Y}(\Omega, \mu; \mathbb{R})$.

1c) the restrictions $u'_{n|M}$ converge in measure to $u_{\infty|M}$;

1d) the restrictions $u'_{n|W}$ converge purely weakly to $u_{\infty|W}$.

2) Moreover if the sequence $(|u_n(.)|^2)_n$ is uniformly integrable, for any $A \in \mathcal{B}(\Omega)$,

$$\Big(\big\|1_A[u'_n - u_\infty]\big\|_{L^2(\Omega)}\Big)^2 \longrightarrow \iint_{A\times\mathbb{R}} |\lambda - u_\infty(x)|^2 \, d\nu(x,\lambda) \,. \qquad (1)$$

Then the restrictions $u'_{n|M}$ converge strongly to $u_{\infty|M}$, and for any Borel subset A of W with $|A| > 0$, $\lim_{n\to\infty} \big\|1_A[u'_n - u_\infty]\big\|_{L^2(\Omega)}$ exists and is > 0.

Comments. 1) Once $\mathrm{bar}(\nu_x) = u_\infty(x)$ is known, one can define M by

$$M := \{x : \int_\Omega |\lambda - u_\infty(x)| \, d\nu_x(\lambda) = 0\} \,.$$

So the measurability of M and W are easy to check.

2) The results extend to L^p with $p \in]1, +\infty[$.

3) In L^1 one has to assume uniform integrability from the very beginning (or to use the biting Lemma (see [SV2]).

4) These results extend to abstract measured space $(\Omega, \mathcal{F}, \mu)$.

5) Even if u_n converges weakly to u_∞ the necessity of extraction of a subsequence to get a convergent sequence of Young measures is well known. For examples see [Va2–3] and the book [Rb] (specially for its fine figures).

Example. Let $\Omega =]0, 1[$ and $u_n = n^{1/2}\, 1_{]0,\frac{1}{n}[}$. Then $\nu_x = \delta_0$. Moreover the convergence $u_n \xrightarrow{\sigma(L^2,L^2)} u_\infty = 0$ holds because $\|u_n\|_2 = 1$ and $\|u_n\|_1 = n^{-1/2} \to 0$. Uniform integrability of the $|u_n(.)|^2$ is lacking and formula (1) does not hold for $A = \Omega$: the left member equals 1, the second 0.

Proof. 1) From the hypothesis $(u_n)_n$ is also $\|.\|_{L^1}$-bounded and the Markov inequality implies

$$\forall \varepsilon > 0, \quad \exists K \in [0, +\infty[\text{ such that } \forall n, \ |\{x : |u_n(x)| \geq K\}| \leq \varepsilon \,.$$

Thus the sequence $(\nu^n)_n$ of the Young measures associated to u_n satisfies the Prokhorov condition and there exists a narrow convergent subsequence $\nu'^n \to \nu$. Thanks to a lower semi-continuity result (which extends to Young measures a part of the portmanteau theorem: see [Bd1], [Va3, Lemma 5]), the following calculus holds:

$$\int_\Omega \Big[\int_\mathbb{R} |\lambda| \, d\nu_x(d\lambda)\Big] \, dx = \iint_{\Omega\times\mathbb{R}} |\lambda| \, d\nu \leq \underline{\lim} \iint_{\Omega\times\mathbb{R}} |\lambda| \, d\nu'^n = \underline{\lim} \, \|u'_n\|_1 < +\infty \,.$$

Thus ν_x is of first order for a.e. x in Ω.

Then the following convergence

$$\iint_{\Omega\times\mathbb{R}} \psi\, d\nu'^n \longrightarrow \iint_{\Omega\times\mathbb{R}} \psi\, d\nu \tag{2}$$

holds not only for bounded Carathodory integrands but even more for ψ an unbounded Carathodory integrand as soon as the sequence $\Big(\psi(.,u'_n(.))\Big)_n$ is uniformly integrable (see [Bd1], [Bd4], [Va2–3]).

Now we prove $u'_n \xrightarrow{\sigma(L^1,L^\infty)} [x \mapsto \mathrm{bar}(\nu_x)] =: u_\infty$. By the de La Vallée Poussin criterion, the boundedness in L^2 implies that $(u_n)_n$ is uniformly integrable. Let $p \in L^\infty$ and set $\psi(x,\lambda) = p(x)\lambda$. Then $\iint_{\Omega\times\mathbb{R}} \psi\, d\nu'^n = \langle p, u'_n\rangle$ and $\iint_{\Omega\times\mathbb{R}} \psi\, d\nu = \langle p, \mathrm{bar}(\nu_.)\rangle = \langle p, u_\infty\rangle$. So (2) gives the expected $\sigma(L^1,L^\infty)$-convergence.

Then $u_\infty \in L^2$ follows from the Jensen inequality and from the lower semi-continuity result already invoked:

$$\begin{aligned}\int_\Omega |u_\infty(x)|^2\, dx = \int_\Omega \Big[\int_\mathbb{R} \lambda\, d\nu_x(d\lambda)\Big]^2 dx \le \int_\Omega \Big[\int_\mathbb{R} |\lambda|^2\, d\nu_x(d\lambda)\Big]\, dx \\ \le \underline{\lim} \iint_{\Omega\times\mathbb{R}} |\lambda|^2\, d\nu'^n = \underline{\lim}\, \big(\|u'_n\|_2\big)^2 < +\infty\,.\end{aligned}$$

Now 1b) follows from the boundedness of $(u_n)_n$ in L^2 and from the denseness of L^∞ in L^2. Thus 1a) and 1b) are proved.

It is time to observe that two possibilities occur: either ν is carried by a graph or not. In the first case, necessarily $M = \Omega$, $\nu_x = \delta_{u_\infty(x)}$ and (cf. [Va2, Prop.6], [Va3, Prop.1]) $u'_n \xrightarrow{\text{meas}} u_\infty$. The second case when $|W| > 0$ is far more interesting.

Statement 1c) is a particular case of the foregoing observations. And 1d) follows from the fact that, for any $A \subset W$, and any further subsequence $(u''_n)_n$ the limit Young measure remains the same and still have the property: [for almost every x in A, ν_x is not a Dirac mass].

2) We suppose the sequence $(|u_n(.)|^2)_n$ uniformly integrable. Let $A \in \mathcal{B}(\Omega)$ and $\psi(x,\lambda) := 1_A(x)\,|\lambda - u_\infty(x)|^2$. Then applying (2) to ψ gives (1). For $A = M$ one gets 0 as limit in (1) and for $A \subset W$, $|A| > 0$, the limit is > 0.

An alternative proof of the strong convergence on M follows from Lebesgue-Vitali's theorem: one has uniform integrability and $|u'_n - u_\infty|^2 \xrightarrow{\text{meas}} 0$. □

Illustrations. Recall a general fact: a.e. ν_x is carried by $\mathrm{Ls}(u'_n(x))$, the set of limit points of the sequence $(u'_n(x))_n$ (see [Bd1], [Va2, Prop.5]). Since $\mathrm{Ls}(u'_n(x))$ is closed the sequence $(u'_n(x))_n$ has at least as limit points all the points of the support of ν_x. They are several as soon as ν_x is not a Dirac measure.

We will describe two different, almost opposite, behaviors: the one of periodical sequences (and in next section the modulated periodical case) which have a regular structure and the behavior of i.i.d. random variables. Both rarely meet. This happens with the Rademacher functions (see 3) below). Surely the random case is often chaotic.

1) The periodical case is (recall $Y = [0,1[^N$): there exists a function $w \in L^2(Y)$ such that $u_n(x) = w(\mathrm{frac}(nx))$. Then

$$u_n \xrightarrow{\sigma(L^2,L^2)} \Big(\int_Y w(y)\,dy\Big)1_\Omega\,.$$

This follows from a well known result in homogenization ([Su], [BM], [Da]) quoted at the beginning of next section. With the same technique one can prove that the Young measures ν^n converge narrowly to $\nu = dx \otimes \theta$ where θ is the image of dy by w. This is also a consequence of the two-scale study: by Part 3) of Theorem 4 below the two-scale Young measures converge to σ with $\sigma_{(x,y)} = \delta_{w(y)}$ and by a result of W. E (reproduced in the remark of [Va5] pages 160–161) ν_x is given by the following weak integral of measures

$$\nu_x = \int_Y \sigma_{(x,y)}\,dy = \int_Y \delta_{w(y)}\,dy = \theta\,.$$

Note that different functions w_1 and w_2 may give the same Young measure (for example with $N = 1$, $w_1(y) = y$ and $w_2(y) = 1 - 2|y - 1/2|$ both give for θ the Lebesgue measure on $[0,1]$). The two-scale convergence method will permit to recover the "pattern" w and even more, in the modulated periodical case, the "mother" if it exists (see formula (5)).

2a) Suppose $|\Omega| = 1$, let P be a probability law of order 2 on $\mathbb{R}$. There always exists a sequence $(u_n)_{n\geq 1}$ of i.i.d. random variables of law P on the probability space $(\Omega, \mathcal{B}(\Omega), dx)$. Indeed there exists a general isomorphism theorem (P.R. Halmos [H, Th.C Sec.41 p.173]; see also R.M. Dudley [Du, Comment on §8.2 p.215]): all nonatomic countably generated probability spaces are isomorphic. One could even use $(]0,1[\,, \mathcal{B}(]0,1[), dx)$.

Proposition 2. Using the foregoing hypotheses the Young measures ν^n converge narrowly to $dx \otimes P$ and u_n converges weakly to $\mathbf{E}(u_1)1_\Omega$.

Proof. By [Va3, Proof of Th.5 p.362] it suffices to prove that for any $\varphi \in \mathcal{C}_c(\mathbb{R})$ and any $p \in L^\infty(\Omega, \mathcal{B}(\Omega), dx)$,

$$\begin{aligned}\iint_{\Omega\times\mathbb{R}} p(x)\varphi(\lambda)\,d\nu^n(x,\lambda) &\longrightarrow \iint_{\Omega\times\mathbb{R}} p(x)\varphi(\lambda)\,d(\mathcal{H}^N \otimes P)(x,\lambda)\\ &= \int_\Omega p(x)\Big[\int_{\mathbb{R}} \varphi\,dP\Big]\,dx\,,\end{aligned} \tag{3}$$

or equivalently

$$\int_\Omega p(x)\varphi(u_n(x))\,dx \longrightarrow \int_\Omega p(x)\Big[\int_{\mathbb{R}} \varphi\,dP\Big]\,dx\,, \tag{3'}$$

that is $\varphi \circ u_n$ converges $\sigma(L^1, L^\infty)$ to $[\int_{\mathbb{R}} \varphi\, dP]1_\Omega$. Let $\mathcal{F}_m$ denote the σ-field of subsets of Ω generated by $u_1, \ldots, u_m$ and $\mathcal{F}$ the σ-field generated by $\cup_{m \geq 1} \mathcal{F}_m$. Let $A \in \mathcal{F}_m$. Then, for $n > m$, by independance

$$\int_\Omega 1_A(x)\varphi(u_n(x))\, dx = \mathbf{E}(1_A)\,\mathbf{E}(\varphi \circ u_n) = \int_\Omega 1_A(x)\Big[\int_{\mathbb{R}} \varphi\, dP\Big]\, dx\,,$$

hence for any $A \in \cup_{m \geq 1} \mathcal{F}_m$

$$\lim_{n \to \infty} \int_\Omega 1_A(x)\varphi(u_n(x))\, dx = \int_\Omega 1_A(x)\Big[\int_{\mathbb{R}} \varphi\, dP\Big]\, dx\,. \tag{4}$$

Since $\cup_{m \geq 1} \mathcal{F}_m$ is dense with respect to the Nikodým metric in $\mathcal{F}$, (4) still holds for $A \in \mathcal{F}$ and by linear combinations and uniform limits of steps functions (3') is true when $p \in L^\infty(\Omega, \mathcal{F}, dx)$. Then classically, for any $p \in L^\infty(\Omega, \mathcal{B}(\Omega), dx)$,

$$\begin{aligned}\int_\Omega p(x)\varphi(u_n(x))\, dx &= \int_\Omega [\mathbf{E}^{\mathcal{F}}(p)](x)\varphi(u_n(x))\, dx \\ &\longrightarrow \int_\Omega [\mathbf{E}^{\mathcal{F}}(p)](x)\Big[\int_{\mathbb{R}} \varphi\, dP\Big]\, dx = \int_\Omega p(x)\Big[\int_{\mathbb{R}} \varphi\, dP\Big]\, dx\,.\end{aligned}$$

Now the weak convergence $u_n \xrightarrow{\sigma(L^2, L^2)} \mathbf{E}(u_1)1_\Omega$ follows from Part 1b) of Theorem 1 or can be proved (more easily) directly. □

2b) For almost every x the sequence $u_n(x)$ has a wandering behavior. It visits (in the ergodic sense) the support of P. Indeed let λ belonging to the support of P and $\varepsilon > 0$. Then $P(]\lambda - \varepsilon, \lambda + \varepsilon[)$ is > 0 and by the Borel-Cantelli Lemma, for almost every x, $u_n(x) \in]\lambda - \varepsilon, \lambda + \varepsilon[$ for infinitely many n. This holds with the same negligible set if $]\lambda - \varepsilon, \lambda + \varepsilon[$ runs through a countable set of open intervals which meet the support of P.

3) Let us give some observations about the Rademacher functions defined by $u_n(x) = \mathrm{sgn}[\sin(2^n \pi x)]$ $(n \geq 1)$. For any subsequence $(u'_n)_n$, the law of large numbers holds: $\frac{1}{n}\sum_{i=1}^n u'_i(x) \to 0$ a.e. But for any fixed subsequence $(u^0_n)_n$ there does not exist a fixed negligible set N such that for any subsequence $(u'_n)_n$ of $(u^0_n)_n$ $\frac{1}{n}\sum_{i=1}^n u'_i(x) \to 0$ holds for any $x \in \Omega \backslash N$. (This shows that in the Komlós theorem [K] the negligible set depends on the subsequence.) Indeed suppose that such an N exists and let $x \in \Omega \backslash N$. Note that $u^0_n(x)$ remains in a compact subset of $\mathbb{R}$ (to be precise in $\{-1, 0, 1\}$) and for any limit point ξ of the sequence $(u^0_n(x))_n$ there would exist a subsequence $(u'_n(x))_n$ converging to ξ. Thus $\frac{1}{n}\sum_{i=1}^n u'_i(x)$ would converge to ξ. Necessarily $\xi = 0$, so the whole sequence $(u^0_n(x))_n$ would converge to 0.

Here is a further easy remark about Rademacher functions: there is not any subsequence $(u'_n)_n$ convergent almost everywhere (but there exist generalized sub-sequences pointwisely convergent!). Indeed, by Lebesgue's theorem it should converge strongly in L^1, but not any subsequence is Cauchy in the L^1-norm.

3. Application of two-scale Young measures

In the following we look for a function (a *mother*) if it exists which generates the oscillating sequence $(u_n)_n$ or at least for a best approximation.

Definition. *A real function $\bar{u}$ on $\Omega \times Y$ (recall $Y = [0,1[^N$) is a mother for the sequence $(u_n)_n$ if*

$$\forall n, \quad u_n(x) \overset{\text{a.e.}}{=} \bar{u}(x, \text{frac}(nx)) \,. \tag{5}$$

Example. The easiest case is the following particular case: there exist functions $v_i \in L^2(\Omega)$ (resp. $v_i \in L^\infty(\Omega)$) and $w_i \in L^\infty(Y)$ (resp. $w_i \in L^2(Y)$) such that $\bar{u}(x,y) = \sum_{i=1}^k v_i(x)\, w_i(y)$ and

$$u_n(x) = \sum_{i=1}^k v_i(x)\, w_i(\text{frac}(nx)) \,.$$

Then $u_n \xrightarrow{\sigma(L^2,L^2)} \sum_{i=1}^k \Big(\int_Y w_i(y)\, dy\Big) v_i$. Indeed for any $p \in L^2(\Omega)$ and any i,

$$\begin{aligned} \int_\Omega p(x)\, v_i(x)\, w_i(\text{frac}(nx))\, dx &\longrightarrow \int_\Omega p(x)\, v_i(x) \Big[\Big(\int_Y w_i(y)\, dy\Big) 1_\Omega(x)\Big]\, dx \\ &= \Big\langle p, \Big(\int_Y w_i(y)\, dy\Big)\, v_i \Big\rangle \,. \end{aligned}$$

because, as well known in homogenization ([Su], [BM], [Da]; the statement is reproduced in [Va5, Prop.1]),

$$w_i(\text{frac}(n\,.)) \xrightarrow{\sigma(L^\infty,L^1)} \Big(\int_Y w_i(y)\, dy\Big)\, 1_\Omega$$

$$(\text{resp. } w_i(\text{frac}(n\,.)) \xrightarrow{\sigma(L^2,L^2)} \Big(\int_Y w_i(y)\, dy\Big)\, 1_\Omega \,) \,.$$

Comment. Given a bounded sequence in $L^2(\Omega)$ the general phenomenon is, up to the extraction of a subsequence, that a part of the oscillations comes from a *modulated* $(\frac{1}{n}\,Y)$*-periodical* [2] behavior as in (5) and another part escape to this modelling. This will be stated precisely in Theorem 5: the notion of two-scale Young measures will give a kind of orthogonal decomposition with quantitative results.

---

[2] In the litterature a strictly positive number ε is used instead of $\frac{1}{n}$. This is better, but for our theoritical exposition n is enough.

Definition. *The two-scale Young measure associated to* $u_n \in L^2(\Omega)$ *is the measure* σ^n *on* $\Omega \times Y \times \mathbb{R}$ *which is the image of* dx *by*

$$x \mapsto \big(x, \operatorname{frac}(\underset{\uparrow}{n}x), u_n(x)\big). \tag{6}$$

In some formulas we will omit frac *when it is clear that* nx *shall belong to* Y*; this will happen in Theorem 5.*

Remarks. 1) The n (ε^{-1} in the literature) above the vertical arrow needs a kind of tuning with (the oscillations of) u_n. For example, if $u_n(x) = \sin(2\pi n^2 x)$, one can consider that the mother is $(x,y) \mapsto \sin(2\pi y)$, but with n^2 instead of n. Here the good function in (6) is $x \mapsto \big(x, \operatorname{frac}(n^2 x), u_n(x)\big)$.

2) The good topology on Y is the one of the torus (the torus is the quotient space $\mathbb{R}^N/\mathbb{Z}^N$). For example when $N = 1$, any point excepted 0 has the same neighborhoods as in the usual topology induced by $\mathbb{R}$ and a basis of neighborhoods of 0 is formed by the sets $[0,\delta[\,\cup\,]1-\delta,1[$ ($\delta > 0$). Thus a continuous function on Y extends in a 1-periodic continuous function on $\mathbb{R}$ and functions such as $(y,\lambda) \mapsto \lambda^2$ are inf-compact on $Y \times \mathbb{R}$ (see [Va5] for the usefulness of these observations).

3) The measure σ^n is the classical Young measure associated to

$$\left| \begin{array}{l} x \mapsto (\operatorname{frac}(nx), u_n(x)) \\ \Omega \to Y \times \mathbb{R}. \end{array} \right.$$

Hence we can easily use test functions which are Carathodory on $\Omega \times Y \times \mathbb{R}$ with continuity with respect to the couple of variables (y,λ) in $Y \times \mathbb{R}$. But some relaxation of the continuity assumption on Y is very useful (see [Va5, Prop.5]).

4) Another special feature is that the useful disintegration (of $\lim_n \sigma'^n$) will be relative to the factorization $(\Omega \times Y) \times \mathbb{R}$.

Definition. *A function* $z : \Omega \times Y \to \mathbb{R}$ *is said to be admissible if it satisfies: for any* $\varepsilon > 0$, $\exists K$ *a compact subset of* Ω *and* $\exists Q$ *a compact subset of* Y *satisfying*

$$\textbf{(Adm)} \qquad \begin{cases} |\Omega \backslash K| \le \varepsilon \\ |Y \backslash Q| \le \varepsilon \\ z_{|K \times Q} \quad \textit{is continuous.} \end{cases}$$

By $\mathbf{Adm}^2(\Omega; Y)$ *we denote the set of all admissible functions satisfying* $\exists \alpha \in L^2(\Omega)$ *such that for all* $(x,y) \in (\Omega \backslash N) \times Y$ *where* N *is* dx*-negligible,* $|z(x,y)| \le \alpha(x)$.

Remarks. Any measurable function z on $\Omega \times Y$ satisfies the Lusin property: $\forall \varepsilon > 0$, $\exists H$ a compact subset of $\Omega \times Y$ such that $|(\Omega \times Y) \backslash H| \leq \varepsilon$ and $z_{|H}$ continuous. The property (**Adm**) is strictly stronger. Sufficient conditions for z to be admissible are (these are the three main cases considered in [Al1]):

1) z has the form $z(x, y) = v(x)w(y)$ where v and w are Borel (or Lebesgue measurable) functions respectively defined on Ω and Y. This follows from the Lusin property applied to v and w. This still holds for a finite sum such as $z(x, y) = \sum_{i=1}^{k} v_i(x)w_i(y)$.

2) z is Carathéodory continuous on Y. Indeed by the Scorza Dragoni theorem [SD], [ET, VIII.1.3 p.218], there exists a compact Q_δ satisfying $|\Omega \backslash Q_\delta| \leq \delta$ and $z|_{Q_\delta \times Y}$ is continuous.

3) z is Carathéodory continuous on Ω (still Scorza Dragoni).

There is a difficulty if one wants use as a mother a function [3] $\bar{u} \in \mathcal{L}^2(\Omega \times Y)$: the values of the u_n depends only on the values of $\bar{u}$ on $\Delta := \cup_{n \in \mathbb{N}^*} \mathrm{gr}(x \mapsto \mathrm{frac}(nx))$ which is a negligible subset of $\Omega \times Y$. Thanks to the following, this difficulty disappears with admissible functions.

Proposition 3 ([Va5, Prop.3]). If $\bar{u}$ and $\tilde{u}$ are admissible and $\bar{u}(x, y) = \tilde{u}(x, y)$ $dx \otimes dy$-a.e., there exist negligible sets $N \subset \Omega$ and $Z \subset Y$ such that $\bar{u}$ and $\tilde{u}$ coincide on $(\Omega \backslash N) \times (Y \backslash Z)$. Then for any $\varepsilon > 0$, $\bar{u}(x, \frac{x}{\varepsilon}) = \tilde{u}(x, \frac{x}{\varepsilon})$ dx-a.e.

Theorem 4. Let $(u_n)_n$ be a bounded sequence in $L^2(\Omega)$. There exists a narrowly convergent subsequence $(\sigma'^n)_n$. Its limit σ has necessarily as projection on $\Omega \times Y$ the Lebesgue measure $dx \otimes dy = \mathcal{H}^{2N}\lfloor_{\Omega \times Y}$. Then σ and its disintegration $(\sigma_{(x,y)})_{(x,y) \in \Omega \times Y}$ with respect to the product $(\Omega \times Y) \times \mathbb{R}$ have the properties:

1) for $dx \otimes dy$-almost every (x, y), $\sigma_{(x,y)}$ is of first order, that is

$$\int_{\mathbb{R}} |\lambda| \, d\sigma_{(x,y)}(\lambda) < +\infty \, .$$

Setting $\hat{u}(x, y) = \int_{\mathbb{R}} \lambda \, d\sigma_{(x,y)}(\lambda)$, one has $\hat{u} \in L^2(\Omega \times Y)$ and

$$\forall \psi \in \mathbf{Adm}^2(\Omega; Y), \quad \int_{\Omega} u'_n(x) \, \psi(x, nx) \, dx \longrightarrow \iint_{\Omega \times Y} \hat{u}(x, y) \, \psi(x, y) \, dx dy \tag{7}$$

2) $u_\infty(x) = \int_Y \hat{u}(x, y) \, dy$ defines a function belonging to $L^2(\Omega)$ and u'_n converges weakly to u_∞ in $L^2(\Omega)$.

---

[3] Classically L^2 denotes the quotient space and $\mathcal{L}^2$ the "prequotient" space. Usually everybody writes L^2 but here the difference is of importance.

3) Moreover if u_n is generated by a mother $\bar{u} \in \mathbf{Adm}^2(\Omega; Y)$, that is $u_n(x) = \bar{u}(x, \text{frac}(nx))$, the whole sequence $(\sigma^n)_n$ converges, its limit is the image of $dx \otimes dy$ by $(x, y) \mapsto (x, y, \bar{u}(x, y))$ (equivalently $\sigma_{(x,y)} = \delta_{\bar{u}(x,y)}$) and $\hat{u} = \bar{u}$.

Remarks. 1) By a denseness result proved by G. Allaire [Al1 Lemma 1.13 p.1490], any $\hat{u} \in L^2(\Omega \times Y)$ can be obtained as the limit in the sense of (7) of some sequence $(u_n)_n$. Our proofs of some consequences hold only if $\hat{u}$ belongs to $\mathbf{Adm}^2(\Omega; Y)$.

2) Formula (7) is the characterization of $\hat{u}$ in the two-scale convergence method. The first proof of the existence of $\hat{u}$ is due to Nguetseng [Ng]. This can be proved directly at least if the class of test functions is not too large; for $L^2(\Omega; \mathcal{C}(Y))$ see [Al1]; for $\mathcal{C}_c(\Omega \times Y)$ this is easier, see [Va5, Th.4]: the left hand-side of (7) is nothing else but[4] $\iint_{\Omega \times Y} \psi \, d\tau^n$ where τ^n is the image of the measure $u_n(.)dx$ by $x \mapsto (x, \text{frac}(nx))$. So it remains to prove that any measure τ which is a $\sigma(\mathcal{M}^b(\Omega \times Y), \mathcal{C}_c(\Omega \times Y))$ limit point of the sequence $(\tau^n)_n$ has a density belonging to $L^2(\Omega \times Y)$. This is rather straightforward.

3) The two-scale convergence in stochastic problems has been studied in [BMW] (there two-scale Young measures are still usefull: see in [Va4] section 7 whose idea is due to G. Michaille).

Some ideas of the proof. The Prokhorov condition is satisfied thanks to the L^2-boundedness. So there exists a narrowly convergent subsequence $(\sigma'^n)_n$. The projection θ^n of σ^n on $\Omega \times Y$ is nothing else but the image of dx by $x \mapsto (x, \text{frac}(nx))$. It converges to $dx \otimes dy$ (this comes from the periodicity of $x \mapsto \text{frac}(nx)$, see for example [Va5, Prop.2]). For the properties stated in 1) and 2) see [Va5, Th.8]. For 3) see [Va5, Prop.6]; here the key argument of the proof is easy when $\bar{u}$ is continuous in the couple of variables (x, y): σ^n is carried by $\text{gr}(\bar{u})$ which is a closed subset of $\Omega \times Y \times \mathbb{R}$, so any narrow limit point of $(\sigma^n)_n$ is still carried by $\text{gr}(\bar{u})$. □

Since, for any $(x, y) \in \Omega \times Y$, $\hat{u}(x, y)$ is the mean of $\sigma_{(x,y)}$, Huygens' formula says:

$$\forall r \in \mathbb{R}, \quad \int_{\mathbb{R}} |\lambda - r|^2 \, d\sigma_{(x,y)}(\lambda) = \int_{\mathbb{R}} |\lambda - \hat{u}(x, y)|^2 \, d\sigma_{(x,y)}(\lambda) + |\hat{u}(x, y) - r|^2 \tag{8}$$

In the following theorem $||.||_2$ means $||.||_{L^2(\Omega)}$.

Theorem 5. Suppose that the sequence $(|u_n(.)|^2)_n$ is uniformly integrable and that $\hat{u}$ belongs to $\mathbf{Adm}^2(\Omega; Y)$. The subsequence under consideration satisfies:

$$\lim_{n\to\infty} (||u'_n - u_\infty||_2)^2 = \lim_{n\to\infty} (||u_n - \hat{u}(., n\,.)||_2)^2 + \lim_{n\to\infty} (||\hat{u}(., n\,.) - u_\infty||_2)^2 \tag{9}$$

[4] For the subsequence under consideration.

where

$$\lim_{n\to\infty}(\|u_n - \hat{u}(.,n\,.)\|_2)^2 = \iiint_{\Omega\times Y\times\mathbb{R}} |\lambda - \hat{u}(x,y)|^2\, d\sigma(x,y,\lambda)$$
$$= \iint_{\Omega\times Y}\Big[\int_{\mathbb{R}} |\lambda - \hat{u}(x,y)|^2\, d\sigma_{(x,y)}(\lambda)\Big]\, dxdy \qquad (10)$$

and

$$\lim_{n\to\infty}(\|\hat{u}(.,n\,.) - u_\infty\|_2)^2 = \iint_{\Omega\times Y} |\hat{u}(x,y) - u_\infty(x)|^2\, dxdy \qquad (11)$$

Moreover for any $\tilde{u} \in \mathbf{Adm}^2(\Omega, dx; Y)$,

$$\lim_{n\to\infty}(\|u'_n - \tilde{u}(.,n\,.)\|_2)^2 = \iiint_{\Omega\times Y\times\mathbb{R}} |\lambda - \tilde{u}(x,y)|^2\, d\sigma(x,y,\lambda)$$
$$= \iint_{\Omega\times Y}\Big[\int_{\mathbb{R}} |\lambda - \tilde{u}(x,y)|^2\, d\sigma_{(x,y)}(\lambda)\Big]\, dxdy \qquad (10')$$

and this quantity is minimum when $\tilde{u} = \hat{u}$.

Remark. These formulas can be localized. Formula (10') localizes in the following way: $\forall A \in \mathcal{B}(\Omega)$,

$$\lim_{n\to\infty}(\|1_A[u'_n - \tilde{u}(.,n\,.)]\|_2)^2 = \iiint_{A\times Y\times\mathbb{R}} |\lambda - \tilde{u}(x,y)|^2\, d\sigma(x,y,\lambda)\,. \quad (10'')$$

Proof. 1) We begin by (10"). Set $\Psi(x,y,\lambda) = 1_A(x)|\lambda - \tilde{u}(x,y)|^2$. Since $\tilde{u}$ is admissible Ψ is not too much different from a Carathodory integrand on $\Omega \times (Y \times \mathbb{R})$ and Proposition 5 of [Va5] applies:

$$\iint_{\Omega\times Y\times\mathbb{R}} \Psi\, d\sigma = \lim_{n\to\infty} \iiint_{\Omega\times Y\times\mathbb{R}} \Psi\, d\sigma'^n\,.$$

So (10") follows from

$$\iiint_{\Omega\times Y\times\mathbb{R}} \Psi\, d\sigma^n = \int_\Omega 1_A(x)|u_n(x) - \tilde{u}(x,nx)|^2\, dx\,.$$

Hence (10') holds. Now by Huygens' formula, $\int_{\mathbb{R}} |\lambda - \tilde{u}(x,y)|^2\, d\sigma_{(x,y)}(\lambda)$ is minimal when $\tilde{u}(x,y)$ is chosen to be $\hat{u}(x,y)$. Formula (10) is the particular case of (10') when $\tilde{u} = \hat{u}$.

2) The measure θ^n on $\Omega \times Y$ image of dx by $x \mapsto (x, nx)$ converges to $dx \otimes dy$ (this was already noticed in the proof of Theorem 4). Set $\psi(x,y) = |\hat{u}(x,y) - u_\infty(x)|^2$. Now formula (11) follows from the uniform integrability of the functions $|\hat{u}(.,n\,.) - u_\infty(.)|^2$ and from

$$\iint_{\Omega\times Y} \psi\, d\theta^n = \int_\Omega |\hat{u}(x,nx) - u_\infty(x)|^2\, dx\,.$$

3) Formula (9) follows easily from (10') (with $\tilde{u}(x,y) = u_\infty(x)$), (8) (with $r = u_\infty(x)$), (10) and (11). □

Remark 3.1. Comments 1) One has 0 in (10) if and only if $\sigma_{(x,y)}$ is the Dirac mass $\delta_{\hat{u}(x,y)}$. Getting 0 in (10) means that the two-scale method recovers quite well the (sub)sequence. When a strictly positive value is obtained, some oscillations escape to this analysis: maybe by a lack of tuning, or by a lack of regular periodicity (think of deformation by a diffeomorphism; see next section), or by more involved oscillations.

2) The strong convergence $u'_n \to u_\infty$ of the subsequence under consideration holds if and only if the right members of (10) and of (11) are 0. Each one can be 0 independently. For example, if $u_n(x) = \sin(2\pi n x_1)$, (10) is 0 and (11) is not. But, if $u_n(x) = \sin(2\pi n^2 x_1)$, with $x \mapsto \big(x, \mathrm{frac}(nx), u_n(x)\big)$ as function in (6), then $\hat{u}(x,y) \equiv 0$, hence (11) is 0 but (10) is not.

4. Further developments

DEFORMATION BY A DIFFEOMORPHISM.

Let $N \geq 2$, $\Omega = \{x \in \mathbb{R}^N : \|x\| < 1\}$ and $u_n(x) = \cos(2\pi n\|x\|)$. In this example there is obviously no standard periodicity (but locally in $\Omega\backslash\{0\}$ one may expect recover by deformation periodicity in Cartesian coordinates: in this line see [Ax]).

MULTI-SCALE CONVERGENCE.

Let $\Omega = \,]-1,1[$ and

$$u_n(x) = \begin{cases} \sin(2\pi n x) & \text{if } x \geq 0 \\ \sin(2\pi n^2 x) & \text{if } x < 0. \end{cases}$$

Here the good method would be the multi-scale convergence of Allaire & Briane [AB] (see also [Bd5]). Keeping $Y = [0,1[$, one has $u_n(x) = \bar{u}(x, nx, n^2 x)$ with $\bar{u} : \Omega \times Y \times Y \to \mathbb{R}$ defined by

$$\bar{u}(x, y_1, y_2) = \begin{cases} \sin(2\pi y_1) & \text{if } x \geq 0 \\ \sin(2\pi y_2) & \text{if } x < 0. \end{cases}$$

Then if $\hat{u}$ is characterized by $\forall \psi \in \mathbf{Adm}^2(\Omega; Y \times Y)$,

$$\int_\Omega u'_n(x)\, \psi(x, nx, n^2 x)\, dx \longrightarrow \iiint_{\Omega \times Y \times Y} \hat{u}(x, y_1, y_2)\, \psi(x, y_1, y_2)\, dx dy_1 dy_2\,,$$

one gets $\hat{u} = \bar{u}$. Three levels are needed: the large scale one (variable x), the first microscopic one (variable y_1) and a further microscopic one (variable y_2).

References

[Ax] Alexandre, R.: Homogénéisation et $\theta - 2$ convergence. C.R. Acad. Sci. Paris **320**,787–790 (1995)

[Al1] Allaire, G.: Homogenization and two-scale convergence. SIAM J. Math. Anal. **23**, 1482–1518 (1992)

[Al2] Allaire, G.: Two-scale convergence and homogenization of periodic structures. In:School on homogenization. Lecture notes of the courses held at ICTP, Trieste, September 6–17, 1993. SISSA Ref. 140/93/M, Trieste, 1993. pp.1–23

[AB] Allaire, G., Briane, M.: Multiscale convergence and reiterated homogenisation. Proc. Royal Soc. Edinburgh **126A**, 297–342 (1996)

[ACV] Amrani, A., Castaing, C., Valadier, M.: Méthodes de troncature appliquées à des problèmes de convergence faible ou forte dans L^1. Arch. Rational Mech. Anal. **117**, 167–191 (1992)

[Bd1] Balder, E.J.: A general approach to lower semicontinuity and lower closure in optimal control theory. SIAM J. Control Optim. **22**, 570–598 (1984)

[Bd2] Balder, E.J.: On weak convergence implying strong convergence in L_1-spaces. Bull. Austral. Math. Soc. **33**, 363–368 (1986)

[Bd3] Balder, E.J.: Generalized equilibrium results for games with incomplete information. Math. Oper. Res. **13-2**, 265–276 (1988)

[Bd4] Balder, E.J.: Lectures on Young measures. Preprint n° 9517, CEREMADE, Université Paris-Dauphine, France 1995

[Bd5] Balder, E.J.: On compactness results for multi-scale convergence. University of Utrecht 1995 (not yet published)

[BGJ] Balder, E.J., Girardi M., Jalby, V.: From weak to strong types of $\mathcal{L}^1_E$-convergence by the Bocce-criterion. Studia Math. **111**, 241–262 (1994)

[BM] Ball, J., Murat, F.: $W^{1,p}$-quasiconvexity and variational problems for multiple integrals. J. Funct. Anal. **58**, 225–253 (1984)

[BMW] Bourgeat, A., Mikelic, A., Wright, S.: Stochastic two-scale convergence in the mean and applications. J. Reine Angew. Math. **456**, 19–51 (1994)

[Ca] Castaing, C.: Convergence faible et sections extrémales. Sém. Anal. Convexe de Montpellier **18**, 2.1–2.18 (1988)

[Da] Dacorogna, B.: Direct Methods in The Calculus of Variations. Springer-Verlag 1989

[Du] Dudley, R.M.: Real analysis and probability. Wadsworth & Brooks/Cole Mathematics Series. California 1989

[E] E, W.: Homogenization of linear and nonlinear transport equations. Comm. Pure Appl. Math. **45**, 301–326 (1992)

[ET] Ekeland, I., Temam, R.: Analyse convexe et problèmes variationnels. Dunod Gauthier-Villars, Paris 1974 (English edition: North-Holland, Amsterdam, 1976)

[H] Halmos, P.R.: Measure Theory (tenth edition). Van Nostrand, Princeton 1965

[K] Komlós, J.: A generalisation of a problem of Steinhaus. Acta Math. Acad. Sci. Hungar. **18**, 217–229 (1967)

[Ng] Nguetseng, G.: A general convergence result for a functional related to the theory of homogenization. SIAM J. Math. Anal. **20**, 608–623 (1989)

[Rb] Roubíček, T: Relaxation in optimization theory and variational calculus. W. de Gruyter, Berlin 1997

[Rz] Rzużuchowski, T.: Strong convergence of selections implied by weak. Bull. Austral. Math. Soc. **39**, 201–214 (1989)

[SV1] Saadoune, M., Valadier M.: Convergence in measure. The Fréchet criterion from local to global. Bull. Polish Acad. Sci. Math. **43**, 47–57 (1995)

[SV2] Saadoune, M., Valadier, M.: Extraction of a "good" subsequence from a bounded sequence of integrable functions. J. Convex Anal. **2**, 345–357 (1995)

[SD] Scorza Dragoni, G.: Un teorema sulle funzioni continue rispetto ad una e misurabili rispetto ad un'altra variabile. Rend. Sem. Mat. Univ. Padova **17**, 102–106 (1948)

[Su] Suquet, P.: Plasticité et homogénéisation. Thèse de Doctorat d'État, Paris 1982

[Ta1] Tartar, L.: Une nouvelle méthode de résolution d'équations aux dérivées partielles non linéaires. In: Journées d'Analyse Non Linéaire (P. Bénilan, J. Robert eds.). Lecture Notes in Math. **665**, pp.228–241 Springer-Verlag 1978

[Ta2] Tartar, L.: Compensated compactness and applications to partial differential equations. In: Nonlinear Analysis and Mechanics: Heriot-Watt Symposium, vol.IV (R.J. Knops ed.). Research Notes in Math. **39**, pp.136–212 Pitman, London 1979

[Va1] Valadier, M.: Différents cas où, grâce à une propriété d'extrémalité, une suite de fonctions intégrables faiblement convergente converge fortement. Sém. Anal. Convexe **19**, 5.1–5.20 (1989)

[Va2] Valadier, M.: Young measures. In:Methods of Nonconvex Analysis (A. Cellina ed.). Lecture Notes in Math. **1446**, pp.152–188 Springer-Verlag 1990

[Va3] Valadier, M.: A course on Young measures. Workshop di Teoria della Misura e Analisi Reale, Grado, September 19–October 2, 1993. Rend. Istit. Mat. Univ. Trieste **26**, suppl., 349–394 (1994)

[Va4] Valadier, M.: Two-scale convergence and Young measures. Prépublication n° 02 (15 pages). Départ. de Math., Université de Montpellier, France 1995

[Va5] Valadier, M.: Admissible functions in two-scale convergence. Portugaliæ Math. **54**, 147–164 (1997)

[Vi] Visintin, A.: Strong convergence results related to strict convexity. Comm. Partial Differential Equations **9**, 439–466 (1984)

[Y] Young, L.C.: Generalized curves and the existence of an attained absolute minimum in the Calculus of Variations. Comptes Rendus de la Société des Sc. et des lettres de Varsovie cl. III **30**, 212–234 (1937)

Adv. Math. Econ. 1, 115–126 (1999)

Advances in
MATHEMATICAL
ECONOMICS

Chaotic solutions in infinite-time horizon linear programming and economic dynamics

Kazuo Nishimura[1] **and Makoto Yano**[2]

[1] Institute of Economic Research, Kyoto University, Sakyo-ku, Kyoto 606-8501, Japan
[2] Department of Economics, Keio University, 2-15-45 Mita, Minato-ku, Tokyo 108-8345, Japan

Received: March 17, 1998

JEL classification: C6

Mathematics Subject Classification (1991): 90A16

1.

In Nishimura and Yano (1996), we demonstrate that chaos may emerge as a solution to a dynamic linear programming (LP) problem. That result is closely related to a result of Nishimura and Yano (1995), which establishes the possibility of chaotic optimal accumulation in a two-sector model of optimal capital accumulation. This study intends to survey those results and explain the basic relationship between them.

Deneckere and Pelikan (1986) and Boldrin and Montrucchio (1986) demonstrate the possibility of chaotic solutions in a dynamic optimization model of discrete time under the assumption that future economic returns (future utilities) are discounted rather heavily (by about 99% to be precise). The farther into the future a particular return is expected to be born, the more heavily the future return tends to be discounted. This fact implies that an inevitable interpretation of the models of Deneckere and Pelikan and Boldrin and Montrucchio is that the length of a unit time interval is fairly long; in any reasonable economic circumstances, the discounting by 99% implies that the length of a unit time interval is at least in the order of decades.

Chaotic dynamics that arises in such a framework implies a rather slow movement of a state variable. Since the work of Deneckere and Pelikan and Boldrin and Montrucchio, therefore, the important open question has been left whether or not chaotic optimal solutions can appear in a model in which future returns are discounted reasonably weakly.

The linear economic framework of Nishimura and Yano (1995,1996) has been useful for providing positive answers to this open question by revealing that chaotic solutions may appear even if future returns are discounted arbitrarily weakly. In this study, we intend to survey the properties that are known in that framework and to present some questions that are left unsolved.

In what follows, in Section 1 we will discuss the basic motivation of our research in the context of mathematical programming and set up the specific LP problem that we deal with in Section 2. We will state our result in the form of a theorem in Section 3 (a proof of which we provide elsewhere). In Section 4, we will interpret the result in the two-sector model of Nishimura and Yano (1995). In Section 5, we will make concluding remarks concerning the direction of future research.

2.

It has been well-known that dynamic programming can be treated in the standard LP framework by adding time structure. In their classic book, for example, Dorfman, Samuelson and Solow (1958) considered the following LP problem.

$$\begin{cases} \max\limits_{(x_1,\dots,x_T)\geq 0} & \sum_{t=1}^{T} p_t' x_t \\ \text{s. t.} & \begin{pmatrix} A_1 & 0 & 0 & \cdots & 0 \\ -B_1 & A_2 & 0 & \cdots & 0 \\ 0 & -B_2 & A_3 & \cdots & 0 \\ 0 & 0 & -B_3 & \ddots & 0 \\ \vdots & \vdots & \vdots & \ddots & A_T \\ 0 & 0 & 0 & \cdots & -B_T \end{pmatrix} \begin{pmatrix} x_1 \\ x_2 \\ x_3 \\ x_4 \\ \vdots \\ x_T \end{pmatrix} \leq \begin{pmatrix} B_1 x' + d_1 \\ d_2 \\ d_3 \\ d_4 \\ \vdots \\ -A_{T+1} x'' + d_T \end{pmatrix}, \end{cases} \tag{2.1}$$

where t may be thought of as time periods. In this problem, x_t may be thought of as the activity vector in period t. This activity uses $A_t x_t$ as "inputs" in period t. This "input" is constrained by the sum of the outputs produced in the previous period, $B_{t-1}x_{t-1}$, and the endowments d_t. For $t = 2, ..., T-1$, the t-th contraint of the LP problem, $A_t x_t - B_{t-1}x_{t-1} \leq d_t$, stipulates that the "net outputs" do not exceed the given capacity, d_t, in period t. Moreover, the first and the last constraints, $A_1 x_1 \leq B_1 x' + d_1$ and $B_T x_T \geq A_{T+1} x'' + d_T$, are the initial and terminal conditions. Under these constraints, the LP problem stipulates that the value of activities, $\sum_{t=1}^{T} p_t' x_t$, be maximized.

Conceptually, the solution to the dynamic LP problem (2.1) can be characterized by the the Bellman principle (see Bellman, 1957, and Bellman and Kalaba, 1965). To this end, define what is called the value function in the manner of backward induction. That is, define

$$V_T(x_{T-1}) = \max_{x_T} p'_T x_T \text{ s.t. } A_T x_T - B_{t-1} x_{T-1} \leq d_T \text{ and } B_T x_T \geq A_T x'' + d_T, \tag{2.2}$$

and, then, define, inductively for $t = T-1, T-2, ..., 1$,

$$V_t(x_{t-1}) = \max_{x_t} [p'_t x_t + V_{t+1}(x_t)] \text{ s.t. } A_t x_t - B_{t-1} x_{t-1} \leq d_t. \tag{2.3}$$

The sets of solutions to these maximization problems, (2.2) and (2.3) can be written as set valued functions of x_{t-1}, $F_t^T(x_{t-1})$, and the dynamic LP problem (2.1) can be characterized by these set valued functions. That is, if and only if a sequence $x_1, ..., x_T$ is a solution to (2.1), then

$$x_t \in F_t^T(x_{t-1}) \tag{2.4}$$

for $t = 1, 2, ..., T$, where $x_0 = x'$.

Since the system (2.4) is set valued, it does not generate a unique trajectory. Therefore, it is not a dynamical system in the strict sense. Since, however, the system determines the set of optimal trajectories, solving the LP problem (2.1), we may call it a generalized optimal dynamical system.

3.

Our question is whether or not this dynamical system can be a chaotic system. Unless the LP problem (2.1) is modified appropriately, this question is void because chaos is concerned with the behavior of trajectories generated by an autonomous dynamical system and because the generalized optimal dynamical system (2.4) is non-autonomous.

There are two reasons why the system (2.4) is autonomous. The first, and very obvious, reason is that the constraints and the objective function of the dynamic LP problem (2.1) are non-autonomous; if the original structure of the optimization problem is time-dependent, it is not surprizing to observe solutions that behave irregularly over time. The second reason is that the LP problem (2.4) is of finite time horizon. As a result, the period-t value function, $V_t^T(x_{t-1})$, depends on how far the current period is away from the end period, *i.e.*, on $T-t$.

This fact indicates that in order to obtain chaotic solutions, we need to deal with an infinite time horizon problem, which is free from the end period problem. With this consideration, we take the following dynamic LP problem.

$$\left\{ \begin{array}{ll} \max\limits_{(x_1, x_2, ...) \geq 0} & \sum_{t=1}^{\infty} \rho^{t-1} p' x_t \\ \text{s. t.} & \left\{ \begin{pmatrix} A & 0 & 0 & \cdots \\ -B & A & 0 & \cdots \\ 0 & -B & A & \cdots \\ 0 & 0 & -B & \ddots \\ \vdots & \vdots & \vdots & \ddots \end{pmatrix} \begin{pmatrix} x_1 \\ x_2 \\ x_3 \\ x_4 \\ \vdots \end{pmatrix} \leq \begin{pmatrix} Bx' + d \\ d \\ d \\ d \\ \vdots \end{pmatrix} \right. , \end{array} \right. \tag{3.1}$$

where ρ, $0 < \rho < 1$, is the discount factor of future values of activities.

In constructing this LP problem, two modifications are made to the LP problem of the previous section. First, the time-horizon is altered to be infinite. Second, the parameters governing the problem are all time independent except that evaluating activities x_t; $A_t = A$, $B_{t-1} = B$, $d_t = d$ and $p_t = \rho^{t-1}p$. In particular, the vector evaluating activity x_t is uniformly decreasing at a constant rate ρ. For these reasons, the LP problem (3.1) may be said to be quasi-stationary.

The solutions to a quasi-stationary optimization problem follows an autonomous optimal dynamical system. In order to obtain this system, note that the maximum value of the objective function depends on the initial condition x'. Denote this maximum value as $V(x')$. With this value function, the Bellman's principle can be stated as

$$V(x_{t-1}) = \max_{x_t} [px_t + \rho V(x_{t-1})] \text{ s.t. } Ax_t - Bx_{t-1} \leq d. \tag{3.2}$$

The set of solutions to this maximization problem can be expressed as a set valued function of x_{t-1}, $F(x_{t-1})$, with a parameter ρ.

As in the finite-time horizon case above, this set valued function characterizes the set of solutions to the infinite-time horizon LP problem (5). That is, a trajectory $x_1, x_2, ...$ is a solution to (3.1) if and only if it is generated by the following relationship.

$$x_t \in F(x_{t-1}) \tag{3.3}$$

for $t = 1, 2, ...$ and $x_0 = x'$.

4.

The question that we face is whether or not this generalized optimal dynamical system, (3.3), is chaotic. In order to deal with this issue, we need to answer the following two specific questions. (i) Under what condition, the optimal program, (3.3), is in fact a dynamical system of the standard sense, described by a single-valued function instead of the set-valued function? (ii) Under what condition the resulting dynamical system is chaotic? Nishimura and Yano (1996) provide an answer to these questions by constructing an example of the LP problem (3.3) in which the optimal program is chaotic.

In the example of Nishimura and Yano (1996), it is assumed that activity vector x_t and the flow constraint are both two dimensional. That is,

$$x_t = \begin{pmatrix} c_t \\ k_t \end{pmatrix}, \; A = \begin{pmatrix} a_{11} & a_{12} \\ a_{21} & a_{22} \end{pmatrix}, \text{ and } B = \begin{pmatrix} 0 & 0 \\ 0 & 1 \end{pmatrix}.$$

Moreover,

$$p = d = \begin{pmatrix} 1 \\ 0 \end{pmatrix} \text{ and } x' = \begin{pmatrix} 0 \\ k \end{pmatrix}.$$

Given this setting, the LP problem (3.1) can be written as

$$\begin{cases} \max\limits_{(c_1,k_1,c_2,k_2,\dots)\geq 0} & \sum_{t=1}^{\infty} \rho^{t-1} c_t \\ \text{s.t.} & \begin{cases} \text{(i)}\ a_{11}c_t + a_{12}k_t & \leq & 1 \\ \text{(ii)}\ a_{12}c_t + a_{22}k_t & \leq & k_{t-1} \\ \qquad\qquad t & = & 1,2,\dots, \\ \text{(iii)} \qquad\qquad k_0 & = & k \end{cases} \end{cases} \tag{4.1}$$

As is noted above, the solutions to problem (4.1) can be described by a generalized dynamical system. To this end, for each $(k_{t-1}, k_t) \geq 0$, define $c(k_{t-1}, k_t)$ as the maximum value of $c_t \geq 0$ satisfying conditions (i) and (ii) of (4.1).

Proposition 4.1. *For each $k \geq 0$, there is a non-empty subset of R_+, $H(k)$, such that if $(c_1, k_1, c_2, k_2, \dots)$ is a solution to (4.1), then it holds that*

$$k_t \in H(k_{t-1}), t = 1, 2, \dots, \tag{4.2}$$

with $k_0 = x$ and that

$$c_t = c(k_{t-1}, k_t). \tag{4.3}$$

We call system H a generalized optimal dynamical system. If, in particular, H is a function, we call it an optimal dynamical system. In what follows, we will demonstrate that H can in fact be a chaotic optimal dynamical system. For the characterization of chaotic motion, we will use the following result due to Lasota and Yorke (1974) and Li and Yorke (1978).

Proposition 4.2. *Let f be a function on a closed interval I into itself satisfying that it is continuously twice differentible everywhere except one point $b \in I$, and that there is an $\epsilon > 0$ such that $|f'(x)| > 1 + \epsilon$ for any x at which f' exists (expansive and unimodal). Then, there is a unique invariant measure on I, μ, that is ergodic with respect to f and absolutely continuous with respect to the Lebesgue measure.*

The above result implies that almost every trajectory following an expansive and unimodal dynamical system behaves as if it were stochastic. The tent map is a well-known example of an expansive and unimodal system.

Our result implies that if parameter values are suitably chosen, the solutions to the LP problem (4.1) can be described by an optimal dynamical system that is expansive and unimodal. In order to explain this result, in Figure 1, the kinked line OPR illustrates the maximum k_t that satisfies conditions (i) and (ii) of problem (4.1) as well as $(c_t, k_t) \geq 0$, given k_{t-1}. In order to obtain this kinked line, set $c_t = 0$ in conditions (i) and (ii), and see that line OPR is

$$k_t = \min\{\mu k_{t-1}, \mu/(1 + \gamma)\}, \tag{4.4}$$

where $\mu = 1/a_{22}$ and $\gamma = a_{12}/a_{22} - 1$. Denote by $\mathcal{D}$ the region in the non-negative quadrant below and on the kinked line OPR. Then, if and only if

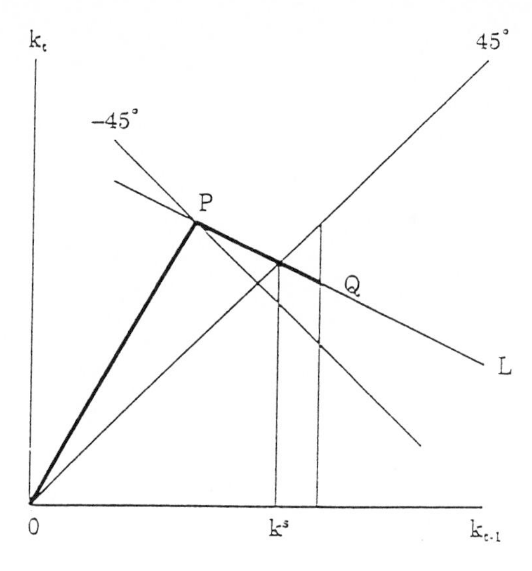

Fig.1

$(k_{t-1}, k_t) \in \mathcal{D}$, there is $c_t \geq 0$ such that conditions (i) and (ii) of problem (4.1) are satisfied.

If conditions (i) and (ii) of problem (4.1) are both satisfied with equality, the value of k_t (as well as that of c_t) is uniquely determined by k_{t-1}. By setting $a_{11}/a_{21} = 1$, this relationship between k_{t-1} and k_t is given by

$$k_t = -(\mu/\gamma)(k_{t-1} - 1). \tag{4.5}$$

If, in particular, $\gamma = a_{12}/a_{22} - 1 > 0$, the graph of equation (4.5) is a negatively-sloped line through point P. In Figure 1, this line is illustrated by line PQ.

The candidate for our chaotic optimal dynamical system is the function the graph of which coincides with the kinked line OPQ; *i.e.*,

$$h(k_{t-1}) = \begin{cases} \mu k_{t-1} & \text{if} \quad 0 \leq k_{t-1} \leq 1/(\gamma+1) \\ -(\mu/\gamma)(k_{t-1} - 1) & \text{if} \quad 1/(\gamma+1) \leq k_{t-1} \leq 1 \end{cases} \tag{4.6}$$

Under the assumption of $\mu/(1+\gamma) \leq 1$, function h maps the unit interval [0,1] into itself. For all the practical purposes, we may restrict h to the closed interval $I = [0, \mu/(1+\gamma)]$ and treat it as a function on I onto itself. Our main result can be stated as follows (see Nishimura and Yano, 1996, for a proof).

Theorem 4.1. *Let* $a_{11}/a_{21} = 1$, $\mu = 1/a_{22}$ *and* $\gamma = a_{12}/a_{22} - 1$. *Moreover, let* h_I *be the function (4.6) restricted to interval* $I = [0, \mu/(\gamma+1)]$. *Suppose that parameters* μ, ρ *and* γ *satisfy*

$$0 < \rho < 1, \quad \rho\mu > 1 \ \ \textit{and} \ \ \mu \leq \gamma + 1. \tag{4.7}$$

Then, on interval I, the generalized optimal dynamical system $H(k)$ coincides with function h_I if one of the following two conditions are satisfied.

$$\textit{Condition A:}\ \ \mu \leq \gamma;$$

$$\textit{Condition B:}\ \gamma < \mu \leq \min\left\{ \frac{\gamma+\sqrt{\gamma^2+4\gamma}}{2}, \frac{-1+\sqrt{1+4\gamma}}{2\rho} \right\}.$$

Function h_I is a tent map. Its graph can be illustrated by the kinked segment OPQ in Figures 1, 2 and 3. Segment OP lies on the ray from the origin with the slope $\mu > 0$ while segement PQ lies on the line through point $(1/(1+\gamma), \mu/(1+\gamma))$ with slope $-\mu/\gamma < 0$.

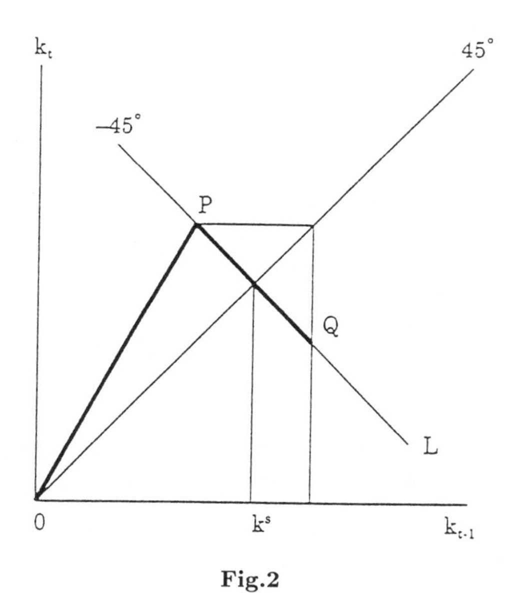

Fig.2

In the case in which Condition A is satisfied, function h_I is unimodal but not expansive; in this case,

$$0 < \rho < 1, \quad \rho\mu > 1 \text{ and } \mu \leq \gamma.$$

If $-\mu/\gamma > -1$, as is shown in Figure 1, optimal dynamical system h_I is globally stable; along any solution to (4.1) with $k > 0$, k_t converges to the non-zero fixed point of system h, k^s. If $-\mu/\gamma = -1$, as is shown in Figure 2, along any solution to (4.1) with $k \neq 0, k^s$, k_t goes into a limit cycle of period 2.

If Condition B is satisfied, as is shown in Figure 3, function h_I is unimodal and expansive. In this sense, by Proposition 1, the optimal dynamical system is chaotic. The existence of chaotic solutions is guaranteed by the next result together with Theorem 1.

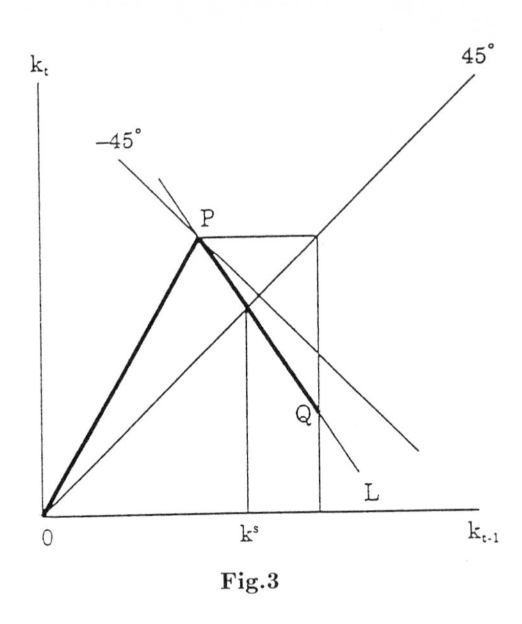

Fig.3

Theorem 4.2. *The set of paramters satisfying (4.7) and Condition B of Theorem 1 at the same time is non-empty.*

An intuitive explanation of our result is as follows. First, notice that if both constraints (i) and (ii) of (4.1) are satisfied with equality, c_t and k_t are uniquely determined for each given k_{t-1}. In Figures 1, 2 and 3, line L captures the relationship between k_{t-1} and k_t in this case, *i.e.*, in the case in which constraints (i) and (ii) are both binding.

This suggests that there may exist cases in which along the optimal solution to (4.1), both of these constraints are binding. In those cases, the graph of the optimal dynamical system lies on line L. Because each variable must be non-negative ($c_t \geq 0$, $k_t \geq 0$ and $k_{t-1} \geq 0$), it may be demonstrated that (k_{t-1}, k_t) cannot lie above point P if (c_t, k_t) satisfies constraints (i) and (ii) for a given $k_{t-1} \geq 0$. In short, only if $k_{t-1} \geq 1/(1+\gamma)$, the graph of the optimal dynamical system can lie on line L.

This suggests that if $k_{t-1} < 1/(1+\gamma)$, the graph of the optimal dynamical system must be at the position closest to line L. This position is given by segment OP; *i.e.*, for a given $k_{t-1} < 1/(1+\gamma)$, the maximum k_t such that (c_t, k_t) satisfies constraints (i) and (ii) together with $c_t \geq 0$ and $k_t \geq 0$ appears on segment OP.

5.

The dynamic LP problem above can be interpreted as a dynamic equilibrium model. In particular, the example (4.1), presented in the previous section, can be thought of as a standard two-sector model with Leontief production functions.

In the case of infinite time horizon, in general, a dynamic equilibrium in a quasi-autonomous economic model can be characterized as a solution to a control problem as follows:

$$\max_{(k_0,k_1,k_2,\ldots)} \sum_{t=1}^{\infty} \rho^{t-1} v(k_{t-1}, k_t) \text{ s.t. } k_0 = k. \tag{5.1}$$

In this optimization problem, k_{t-1} and k_t are thought of as the levels of capital stock at the beginning and end of period t. The utility function $v(k_{t-1}, k_t)$ describes the utility that the economy can achieve in period t in the case in which it starts with stock level k_{t-1} at the beginning of period t and is to achieve stock level k_t at the end of that period. In the optimization (5.1), therefore, it is assumed that in a dynamic equilibrium, a path of capital accumulation is carried out in such a way that the discounted utility sum over the entire time horizon be maximized.

Nishimura and Yano (1995) demonstrates the existence of a chaotic dynamic equilibrium in a special case of this dynamic equilibrium model (5.1). In their example, two goods C and K are assumed to exist. Good C is a pure consumption good. Good K is a produced good that is used in the production processes of both goods. There are two sectors, each specialized in the production of one of the goods. Call them sectors C and K. Each sector uses both good K and labor L as inputs. The input of good K must be carried out one period prior to the period in which output is produced; in this sense, we may think of good K as a capital good. Labor input is made in the same period as output is produced. Good K is not consumed. Sectors C and K have Leontief production functions as follows:

$$c_t = \min\{K_{Ct-1}, L_{Ct}/\alpha\}; \tag{5.2}$$

$$k_t = \mu \min\{K_{Kt-1}, L_{Kt}/\beta\}. \tag{5.3}$$

For $i = C, K$, $K_{it-1} \geq 0$ is sector $i's$ capital input in period $t-1$, and $L_{it} \geq 0$ is sector $i's$ labor input in period t. Moreover, $c_t \geq 0$ and $y_t \geq 0$ are outputs in period t. The aggregate capital input must be less than the output of sector K in the previous period.

$$K_{Ct-1} + K_{Kt-1} \leq k_{t-1}. \tag{5.4}$$

Assume that labor supply is inelastic and time independent. Goods are normalized so that the labor endowment $\bar{n}$ is equal to 1. Since the aggregate labor input cannot exceed the labor endowment, it must hold that

$$L_{Ct} + L_{Kt} \leq 1. \tag{5.5}$$

The consumers' preference is represented by a linear utility function

$$u(c_t) = c_t \geq 0 \tag{5.6}$$

for $c_t \geq 0$ in each period. Given this structure, we may define

$$v(k_{t-1}, k_t) = \max_{(c_t, K_{Ct-1}, K_{Kt-1}, L_{Ct}, L_{Kt}) \geq 0} u(c_t) \text{ s.t. } (5.2)\text{-}(5.6). \tag{5.7}$$

The optimization problem with the specific form of utility function v given by (5.7) is an example of the typical two-sector dynamic equilibrium model.

As is well-known, this optimization problem is a special case of the standard model of a dynamic general equilibrium economy (see, for example, Yano, 1998). In such an equilibrium, the consumers and firms in the economy carry out their respective dynamically optimal economic activities over time, and the market for each good is cleared from the present through the entire future periods. The optimization problem (5.2) is, therefore, called a dynamic equilibrium model.

This dynamic equilibrium model (5.1) with (5.7) can be shown to be equivalent to the dynamic LP problem (4.1). This is because the model has a linear utility function and Leontief production functions. Because of the linear utility function, the objective functions of (5.7) and (4.1) are identical. Given Leontief production functions, moreover, the constraints of (5.7) are basically the same as those of (4.1).

In order to check this fact, think of the case in which $K_{Ct-1} = L_{Ct}/\alpha$ and $K_{Kt-1} = L_{Kt}/\beta$. In this case, given (5.2) and (5.3), it holds that $c_t = K_{Ct-1} = L_{Ct}/\alpha$ and $k_t = \mu K_{Kt-1} = \mu L_{Kt}/\beta$. Therefore, (5.4) and (5.5) can be transformed into

$$c_t + \frac{1}{\mu} k_t \leq k_{t-1}$$

and

$$\alpha c_t + \frac{\beta}{\mu} k_t \leq 1.$$

These constraints are identical to those of (4.1) in the case of $a_{11} = \alpha$, $a_{12} = \beta/\mu$, $a_{21} = 1$ and $a_{22} = 1/\mu$. Since, given (5.7), it may be demonstrated that $K_{Ct-1} = L_{Ct}/\alpha$ and $K_{Kt-1} = L_{Kt}/\beta$ must hold in any solution, the equivalence between (5.1) and (4.1) can be established.

6.

Given the equivalence, under (5.7), between (5.1) and (4.1), Theorem 4.2 demonstrates the possibility that a dynamic equilibrium is governed by a chaotic dynamical system. It was Deneckere and Pelikan (1986) and Boldrin and Montrucchio (1986) that first proved that such a possibility exists. However, their results are based on the case in which the discount factor of future utilities ρ is very small (more precisely, in the order of 10^{-2}).

From the viewpoint of economics, the case in which the discount factor is small is of smaller interest than that in which it is close to 1. This is because the discount factor ρ governs the degree of myopia of consumers with respect to future utilities. In the case in which $\rho = 10^{-2}$, the consumers value the utility that they are to obtain in the next period to be one hundredth of the utility that they have today. In such a case, it must be assumed that the length of a period is quite long (say, about 100 years).

Our results summarized by Theorems 4.1 and 4.2, in contrast, demonstrate that that a dynamic equilibrium can be governed by a chaotic dynamical system for values of discount factor ρ up to 0.5; in other words, the least upper bound of ρ such that (ρ, μ, γ) satisfies (4.7) together with Condition B of Theorem 4.1 is

$$\rho^* = 0.5. \tag{6.1}$$

As discussed above, this upper bound poses a severe limitation on the economic application of chaotic optimal dynamics, for example. Dynamic LP problem (4.1) may be interpreted as a model of capital accumulation, in which c_t and k_t may be thought of as representing, respectively, levels of consumption and capital stock. Under such an interpretation, ρ may be thought of as determining the length of an individual period of the model. It is generally considered that ρ is around 0.95 in economic models in which the length of an individual period is one year. If $\rho < 0.5$, therefore, the length of a period becomes about a half decade. In other words, the economic application of chaotic optimal dynamics is limited to a model in which the length of a single period is assumed to be more than a half decade.

It is important to note that $\rho^* = 0.5$ is not the least upper bound of discount factors with which a chaotic optimal dynamical system can appear. Nishimura and Yano (1995) demonstrate that no matter how close ρ is to 1, it is possible to choose (μ, γ) in such a way that the optimal dynamical system can be chaotic. That result is, however, based on the assumption that point P in Figures 1, 2 and 3 is a cyclical point of dynamical system h_I.

Neither the result of Nishimura and Yano (1995) nor the results reported in this paper provides a complete characterization for dynamical system h_I to be the solution to (4.1). It is an interest question left for future research to derive a necessary and sufficient condition on parameters of the LP problem (4.1) under which dynamical system h_I is optimal. Such a characterization would provide a better understanding on the possibility of chaotic optimal dynamics.

References

1. Bellman, R.: Dynamic Programming. Princeton University Press, Princeton 1957
2. Bellman, R., Kalaba, R.: Dynamic Programming and Modern Control Theory. Academic Press, New York 1965
3. Boldrin, M., Montrucchio, L.: On the indeterminacy of capital accumulation paths. Journal of Economic Theory **40**, 26-39 (1986)
4. Deneckere, R., Pelikan, S.: Competitive chaos. Journal of Economic Theory **40**, 13-25 (1986)
5. Dorfman, R., Samuelson, P., Solow, R.: Linear Programming and Economic Analysis. McGraw-Hill, New York 1958
6. Lasota, A., Yorke, J.: On the existence of invariant measures for piecewise monotonictransformations. Transactions of American Mathematical Society **186**, 481-488 (1974)
7. Li, T.-Y., Yorke, J.: Period three implies chaos. American Mathematical Monthly **82**, 985-992 (1975)
8. _____: Ergodic transformations from an interval into itself. Transactions of American Mathematical Society **235**, 183-192 (1978)
9. Nishimura, K., Yano M.: Non-linear dynamics and chaos in optimal growth: An example. Econometrica **63**, 981-1001 (1995)
10. _____: Chaotic solutions in dynamic linear programming. Chaos, Solitons & Fractals **7**, 1941-1953 (1996)
11. Yano, M.: On the duality of a von Neumann facet and the inefficacy of temporary fiscal policy. Econometrica **66**, 427-451 (1998)

Adv. Math. Econ. 1, 127–133 (1999)

Advances in
MATHEMATICAL
ECONOMICS

Determinacy of monetary equilibria in an economy with no real risk

Shinichi Suda

Department of Economics, Keio University, 2-15-45 Mita, Minato-ku, Tokyo 108-8345, Japan
(e-mail: ssuda@econ.keio.ac.jp)

Received: April 1, 1998

JEL classification: D52, E40

Summary. This paper examines the determinacy of equilibria in an exchange economy with money and a nominal bond where the only source of uncertainty comes from fluctuations in the money supply. Money plays the role of medium of exchange and, through a cash-in-advance constraint, affects the real allocation. We show that the monetary economy can be transformed into a standard Arrow-Debreu economy, and these two economies have the same equilibrium allocations. Applying the theorem on the finiteness of equilibria by Debreu [3], we prove that the set of monetary equilibria is locally unique, generically for every level of money supply.

Key words: Determinacy, monetary equilibrium, incomplete market

1. Introduction

Debreu [3, 4] has emphasized the importance of establishing that generically an economy has a finite set of equilibria. However, in the recent studies of equilibria in economies with incomplete markets and nominal assets (Balasko and Cass [1], Geanakoplos and Mas-Colell [5]), it has been shown that the set of equilibrium allocations of these economies is indeterminant, i.e., the set contains a high-dimensional submanifold of equilibria. Magill and Quinzii [6] is an attempt to reestablish the determinacy of equilibria by explicitly introducing money into the economy.

This paper considers the same problem in a different environment, namely, in economies with no real risk. Here the only source of uncertainty comes from fluctuations in the money supply, and the agents' endowments and preferences are independent of these fluctuations. Using a simple model (a monetary exchange economy), we show that even in this restricted circumstances, generically, the set of monetary equilibria is locally unique. Thus this is considered to be an attempt to restore the determinacy of equilibria in a sunspot economy with incomplete markets and nominal assets (see Cass [2]).

Our model is basically the same as the one in Magill and Quinzii [6, 7], where money plays the role of medium of exchange and, through a cash-in-advance constraint, affects the real allocation. The cash-in-advance constraint is introduced by dividing each period into three subperiods, in the initial subperiod goods being sold in exchange for money, in the second nominal bond being traded among consumers and in the last subperiod money being used to purchase goods. The nominal bond performs the function of a store of value. Since it is the only asset traded in the market, the financial market is incomplete.

We will show that a monetary exchange economy can be transformed into a standard Arrow-Debreu economy, and these two economies have the same equilibrium allocations. Applying the theorem on the finiteness of equilibria by Debreu [3], we then show that the set of monetary equilibria is locally unique, generically for every level of money supply.

The rest of the paper is organized as follows. Section 2 contains the setup of the model and the maintained assumptions throughout the paper. In Section 3 we establish the existence and generic determinacy of monetary equilibria.

2. The model

We consider a monetary exchange economy that lasts for two periods (date 0 and 1). In each period fiat money is used as a medium of exchange and as the unit of account. There are S states of nature in the second period and each of them corresponds to a level of money supply in that period. These states are denoted by the subscript $s = 1, 2, \cdots, S$, while the first period is denoted by $s = 0$. We assume that the fluctuations in the money supply are the only source of uncertainty, so that the consumers' endowments and preferences in the second period are known for sure in the first period.

The consumers know the probabilities of the realization of each state, which are denoted by $\pi_s > 0$, $s = 1, \cdots, S$, $\sum_{s=1}^{S} \pi_s = 1$. For every $s \geq 0$, one physical commodity is traded in the market. It is perishable and cannot be carried over to the next period. For the role of a store of value, a nominal bond is traded in $s = 0$. One unit of this bond cost its owner q units of money in $s = 0$ and pays him 1 unit of money in every $s > 0$.

There are H consumers labeled by the subscript $h = 1, 2, \cdots, H$. $x_h = (x_{h0}, x_{h1}, \cdots, x_{hS}) \in \mathbb{R}^{S+1}_{++}$ and $e_h = (e_{h0}, e_{h1}, \cdots, e_{hS}) \in \mathbb{R}^{S+1}_{++}$ are the consumption and the endowment vectors of consumer h. The holding of bond by consumer h is denoted by θ_h, which is either positive or negative. We will use the notation $x = (x_1, x_2, \cdots, x_H)$ and $\theta = (\theta_1, \theta_2, \cdots, \theta_H)$, and when we say "allocation", we mean x.

The assumptions below are imposed throughout the paper:

Assumption 1. Second period endowments are state-invariant: $e_{h1} = e_{h2} = \cdots = e_{hS}$ for all h.

Assumption 2. Preferences of each consumer h are represented by an expected utility function:

$$u_h : \mathbb{R}^{S+1}_{++} \to \mathbb{R}$$

$$u_h(x_h) = \sum_{s=1}^{S} \pi_s v_h(x_{h0}, x_{hs})$$

where $v_h : \mathbb{R}^2_{++} \to \mathbb{R}$ is, for any h, C^2, differentiably strictly increasing, differentiably strictly concave and has indifference surfaces with closure in $\mathbb{R}^2_{++}$.

These assumptions guarantee that the consumers' endowments and preferences are independent of the second period state, s.

We now describe the structure of the market. Following Magill and Quinzii [6, 7], we decompose each state $s(s = 0, 1, \cdots, S)$ into three subperiods: in the first subperiod, consumers sell their endowment of goods in exchange for money, in the second, consumers transact on the bond market, and in the last subperiod, consumers use their transactions balances to purchase goods. There is an institution that we call the Central Exchange, which performs the basic function of marketing the consumers' endowments. We assume that consumers' endowments are not directly consumable but need to be processed in the Central Exchange, so that in the first subperiod each consumer sells the full amount of his endowment to the Central Exchange. In this way we introduce the cash-in-advance constraint into the model. The Central Exchange also determines the supply of money, M_s, that is injected into the economy in each state $s(s = 0, 1, \cdots, S)$.

In the first subperiod of date 0 each consumer h receive the money income $p_0 e_{h0}$ from the Central Exchange, where p_0 is the price of the commodity at date 0. In the second subperiod the bond is traded among consumers, which leads to a redistribution of the money balances. At the end of this subperiod the money balance of consumer h turns out to be $p_0 e_{h0} - q\theta_h$, and this is used in the third subperiod to buy x_{h0} from the Central Exchange at the price p_0.

At date 1, one of the states $s(s = 1, \cdots, S)$ occurs and the level of money supply is set to be M_s. Again in the first subperiod each consumer h receive the money income $p_s e_{h1}$ in exchange for his endowment, which is sold to the Central Exchange at the price p_s. Then in the second subperiod, consumer h receive the yield from the bond, θ_h. The money balance, $p_s e_{h1} + \theta_h$, is used in the third subperiod to buy x_{hs} from the Central Exchange at the price p_s.

We assume that each consumer correctly anticipate the future prices $(p_1, \cdots, p_S)$, and given $(p, q) = (p_0, p_1, \cdots, p_S; q) \in \mathbb{R}^{S+2}_{++}$ maximize his utility. Thus for $(p, q) \in \mathbb{R}^{S+2}_{++}$, each consumer h solves the problem:

$$\max_{x_h,\theta_h} u_h(x_h)$$
$$\text{subject to} \quad p_0(x_h\hat{0} - e_{h0}) = -q\theta_h$$
$$p_s(x_{hs} - e_{h1}) = \theta_h, \qquad s = 1, \cdots, S$$
$$\text{and} \quad x_h \gg 0.$$

The solution to the problem determines the demand functions $x_{hs}(p,q), s = 0, 1, \cdots, S$, and $\theta_h(p,q)$ for goods and bond.

We can now define the monetary equilibrium.

Definition. A *monetary equilibrium* is a vector of prices, allocations and holdings of bond $(p,q,x,\theta) \in \mathbb{R}^{S+2}_{++} \times \mathbb{R}^{H(S+1)}_{++} \times \mathbb{R}^H$ such that

(a) $x_{hs} = x_{hs}(p,q)$ and $\theta_h = \theta_h(p,q)$ for all s and h,

(b) $\sum_{h=1}^H (x_h - e_h) = 0$,

(c) $\sum_{h=1}^H \theta_h = 0$,

(d) $p_s \sum_{h=1}^H x_{hs} = M_s, \quad s = 0, 1, \cdots, S$. (b) and (c) are the market clearing conditions in the commodity and bond markets. (d) expresses the cash-in-advance constraint.

In the next section we also need the concept of financial equilibrium, which is obtained by dropping the condition (d) in the definition of monetary equilibrium.

Definition. A *financial equilibrium* is a vector of prices, allocations and holdings of bond $(p,q,x,\theta) \in \mathbb{R}^{S+2}_{++} \times \mathbb{R}^{H(S+1)}_{++} \times \mathbb{R}^H$ such that

(a) $x_{hs} = x_{hs}(p,q)$ and $\theta_h = \theta_h(p,q)$ for all s and h,

(b) $\sum_{h=1}^H (x_h - e_h) = 0$,

(c) $\sum_{h=1}^H \theta_h = 0$.

By definition, any monetary equilibrium is a financial equilibrium.

3. Existence and determinacy of monetary equilibria

Our objective is to show the existence and finiteness of monetary equilibria. The monetary exchange economy of the previous section can be summarized by the characteristics of the consumers $(v,e) = (v_1, \cdots, v_H, (e_{10}, e_{11}), \cdots, (e_{H0}, e_{H1}))$ and the monetary policy $M = (M_0, M_1, \cdots, M_S)$. If we fix a profile of preferences $(v_1, \cdots, v_H)$ where each v_h satisfies Assumption 2, we obtain an economy $E(e,M)$ parametrized by the endowments and money supply $(e,M) \in \mathcal{E} \times \mathcal{M}, \mathcal{E} = \mathbb{R}^{2H}_{++}, \mathcal{M} = \mathbb{R}^{S+1}_{++}$, where $\mathcal{E}$ is the endowment space and $\mathcal{M}$ is the monetary policy space.

We prove the existence and finiteness of monetary equilibria by the following theorems.

Theorem 1. The economy $E(e,M)$ has a monetary equilibrium for any $(e,M) \in \mathcal{E} \times \mathcal{M}$.

Theorem 2. For any $M \in \mathcal{M}$, there exists an open subset of full measure in the endowment space, $X_M \subset \mathcal{E}$ such that an economy $E(e, M)$ with $e \in X_M$ has a finite number of monetary equilibrium allocations.

In the rest of the section, we fix a monetary policy $M \in \mathcal{M}$. Without loss of generality we suppose $M_1 = \min(M_1, M_2, \cdots, M_S)$ and define λ by $\lambda = (\lambda_2, \lambda_3, \cdots \lambda_S) = (\frac{M_2}{M_1}, \frac{M_3}{M_1}, \ldots, \frac{M_S}{M_1})$. Moreover for each consumer h, define

$$u_h^\lambda(x_{h0}, x_{h1}) = u_h(x_{h0}, x_{h1}, e_{h1} + \frac{1}{\lambda_2}(x_{h1} - e_{h1}), \cdots, e_{h1} + \frac{1}{\lambda_S}(x_{h1} - e_{h1})).$$

Since $\lambda_s \geq 1$, $s = 2, \cdots, S$, $e_{h1} + \frac{1}{\lambda_s}(x_{h1} - e_{h1}) > 0$ whenever $x_{h1} > 0$, so that u_h^λ is well-defined. It is also easily shown that u_h^λ has the same property as v_h. The following lemma plays the central role in proving Theorem 1 and 2.

Lemma. (i) If $(p, 1, x, \theta)$ is a monetary equilibrium of $E(e, M)$, then $(p_0, p_1, (x_{h0}, x_{h1})_{h=1}^H)$ satisfies the following conditions:
(I) For each consumer h, (x_{h0}, x_{h1}) is the optimal solution to

$$\begin{aligned} &\max_{x_{h0}, x_{h1}} \ u_h^\lambda(x_{h0}, x_{h1}) \\ &\text{subject to} \quad p_0(x_{h0} - e_{h0}) + p_1(x_{h1} - e_{h1}) = 0 \\ &\qquad\qquad \text{and} \quad (x_{h0}, x_{h1}) \gg 0. \end{aligned}$$

(II) $\sum_{h=1}^H (x_{h0} - e_{h0}) = 0$ and $\sum_{h=1}^H (x_{h1} - e_{h1})=0$

(ii) If $(\bar{p}_0, \bar{p}_1, (x_{h0}, x_{h1})_{h=1}^H) \in \mathbb{R}_{++}^2 \times \mathbb{R}_{++}^{2H}$ satisfies (I) and (II), then we can find $(p, q, (x_{h2}, \cdots, x_{hS})_{h=1}^H, \theta) \in \mathbb{R}_{++}^{S+2} \times \mathbb{R}_{++}^{H(S-1)} \times \mathbb{R}^H$ such that (p, q, x, θ) is a monetary equilibrium of $E(e, M)$.

Proof. (i) Since $(p, 1, x, \theta)$ is a monetary equilibrium of $E(e, M)$, for each h, x_h maximizes $u_h(x_h)$ under the constraints

$$\begin{aligned} &p_0(x_{h0} - e_{h0}) + p_1(x_{h1} - e_{h1}) = 0 \\ &p_1(x_{h1} - e_{h1}) - p_s(x_{hs} - e_{h1}) = 0, \quad s = 2, \cdots, S \quad \text{and} \\ &x_h \gg 0. \end{aligned}$$

And since $\lambda_s = \frac{M_s}{M_1} = \frac{p_s}{p_1}$, $s = 2, \cdots, S$ by the equilibrium conditions (b) and (d), (x_{h0}, x_{h1}) solves the optimization problem of (I). The condition (II) is clearly satisfied by the equilibrium condition (b).
(ii) If we define

$$\begin{aligned} &\bar{p}_s = \lambda_s \bar{p}_1, \quad s = 2, \cdots, S, \\ &x_{hs} = e_{h1} + \frac{1}{\lambda_s}(x_{h1} - e_{h1}), \quad h = 1, \cdots, H, \ s = 2, \cdots, S, \text{ and} \\ &\bar{\theta}_h = \bar{p}_1(x_{h1} - e_{h1}), \quad h = 1, \cdots, H, \end{aligned}$$

then $(\bar{p}, 1, x, \bar{\theta})$ is a financial equilibrium of $E(e, M)$. Furthermore by letting

$$p_s = \frac{M_s}{\bar{p}_s \Sigma_h e_{h1}} \bar{p}_s = \frac{M_1}{\bar{p}_1 \Sigma_h e_{h1}} \bar{p}_s, \quad s = 1, \cdots, S,$$

$$\theta_h = \frac{M_1}{\bar{p}_1 \Sigma_h e_{h1}} \bar{\theta}_h, \quad h = 1, \cdots, H,$$

$$q = \frac{M_0 \bar{p}_1 \Sigma_h e_{h1}}{M_1 \bar{p}_0 \Sigma_h e_{h0}} \quad \textit{and}$$

$$p_0 = \bar{p}_0 \frac{M_1}{\bar{p}_1 \Sigma_h e_{h1}} q,$$

we obtain a monetary equilibrium (p, q, x, θ) of $E(e, M)$. QED

After these preparations we proceed to prove Theorem 1 and 2.

Proof of Theorem 1. (I) and (II) characterize a Walrasian equilibrium of a pure exchange economy with $(u_1^\lambda, \cdots, u_H^\lambda, (e_{10}, e_{11}), \cdots, (e_{H0}, e_{H1}))$ being the characteristics of consumers. Then given Assumption 2, we can prove the existence of an equilibrium of this economy by the standard argument. Therefore using Lemma (ii) we can show that a monetary equilibrium of $E(e, M)$ exists. QED

Proof of Theorem 2. Recall that in the beginning of this section, we fixed a monetary policy $M \in \mathcal{M}$, and defined λ by $\lambda = (\frac{M_2}{M_1}, \frac{M_3}{M_1}, \cdots, \frac{M_S}{M_1})$. For any endowment profile $e \in \mathcal{E}$ let $A(e)$ be the projection of the set of equilibrium allocations of $E(e, M)$ into the coordinates of $(x_{h0}, x_{h1})_{h=1}^H$. From our maintained assumptions it is clear that the finiteness of $A(e)$ implies the finiteness of monetary equilibrium allocations. In what follows we will show that $A(e)$ is contained in the set of allocations satisfying (I) and (II) (we call this set $W(e, \lambda)$) and $W(e, \lambda)$ is generically finite.

By Lemma (i) it is clear that the projection of the equilibrium with $q = 1$ is contained in $W(e, \lambda)$. Now, if (p, q, x, θ) is an arbitrary monetary equilibrium, then $(p, 1, x, \theta)$ is a monetary equilibrium of the economy, $E(e, \frac{M_0}{q}, M_1, \cdots, M_S)$. And since the set of equilibrium allocations of $E(e, M_0, M_1, \cdots, M_S)$ and that of $E(e, \frac{M_0}{\alpha}, M_1, \cdots, M_S)$, $(\alpha > 0)$ are the same, and these two monetary policy generate the same λ, we know that the projection of any equilibrium of $E(e, M)$ is contained in $W(e, \lambda)$. Finally by the theorem of Debreu [3] we know the set $W(e, \lambda)$ is generically finite, therefore the set of monetary equilibrium allocations is also generically finite. QED

References

1. Balasko, Y., Cass, D.: The structure of financial equilibrium with exogenous yields: the case of incomplete markets. Econometrica **57**, 135-162 (1989)

2. Cass, D.: Sunspots and incomplete financial markets: the general case. Econ. Theory **2**, 341-358 (1992)
3. Debreu, G.: Economies with a finite set of equilibria. Econometrica **38**, 387-392 (1970)
4. Debreu, G.: The application to economics of differential topology and global analysis: Regular differentiable economies. Amer. Econ. Rev. **66**, 280-287 (1976)
5. Geanakoplos, J., Mas-Colell, A.: Real indeterminacy with financial assets. J. Econ. Theory **47**, 22-38 (1989)
6. Magill, M. Quinzii M.: Real effects of money in general equilibrium. J. Math. Econ. **21**, 301-342 (1992)
7. Magill, M. Quinzii M.: Theory of incomplete markets. vol. 1. Cambridge: MIT Press 1996

Programme: KES/RCME Conference on mathematical analysis in economic theory in honor of Professor Gérard Debreu. October 4-5, 1997, Keio University (Tokyo)

*=speaker

October 4th (Saturday)

Registration (8:20-)

Morning Session

chaired by Kunio Kawamata (Keio University)

I (9:00a.m.-10:00a.m.)

Shinichi Suda (Keio University)

"Determinacy of monetary equilibria in an economy with no real risk"

II (10:10a.m.-11:10a.m.)

Egbert Dierker*(Universität Wien) and Birgit Grodal (Københavns Universitet)

"The price normalization problem in general equilibrium models of imperfect competition"

III (11:20a.m.-12:20p.m.)

Hildegard Dierker*(Technische Universität Wien) and Egbert Dierker (Universität Wien)

"Product differentiation and market power"

lunch

Special Session

chaired by Masao Fukuoka (Kanto Gakuen University)

IV (1:30p.m.-2:30p.m.)

Gérard Debreu (U.C.Berkeley)

"On the use in economic theory of some central results of mathematical analysis"

Afternoon Session

chaired by Kazuo Nishimura (Kyoto University)

V (3:00p.m.-4:00p.m.)

Jean-Michel Grandmont*(CREST) and Laurent Calvet (University of Yale and CERAS)

"Heterogeneity and aggregation"

VI (4:10p.m.-5:10p.m.)

Makoto Yano*(Keio University) and Kazuo Nishimura (Kyoto University)

"Chaotic solutions in infinite-time horizon linear programming and economic dynamics"

Reception Party (6:30p.m.- at Palace Hotel)

October 5th (Sunday)

Morning Session

chaired by Norio Kikuchi(Keio University)

VII (9:00a.m.-10:00a.m.)

Michel Valadier (Université Montpellier II)

"Analysis of the asymptotic distance between oscillating functions and their weak limit in L^2"

VIII (10:10a.m.-11:10a.m.)

Charles Castaing*(Université Montpellier II) and Mohamed Geussous (Université Hassan II Mohamedia)

"Convergences in L^1_X and applications"

IX (11:20a.m.-12:20p.m.)

Tatsuro Ichiishi (Ohio State University)

"*Ex ante* and *interim* contracts signed by the divisions of Chandler's M-form firm"

lunch

Afternoon Session

chaired by Ryozo Miura (Hitotsubashi University)

X (2:00p.m.-3:00p.m)

Andrew McLennan (University of Minnesota)

"On the expected number of Nash equilibria of a normal form game"

XI (3:20p.m.-4:20p.m.)

Hiroshi Shirakawa (Tokyo Institute of Technology)

"Evaluation of the premium yield curve for the default risk"

XII (4:30p.m.-5:30p.m.)

Shigeo Kusuoka (University of Tokyo)

"Replication costs for American derivatives with transaction costs"

Subject Index

Instructions for Authors

A. General

1. Papers submitted for publication will be considered only if they have not been and will not be published elsewhere without permission from the publisher and the Research Center of Mathematical Economics.

2. Every submitted paper will be subject to review. The names of reviewers will not be disclosed to the authors or to anybody not involved in the editorial process.

3. The authors are asked to transfer the copyright to their articles to Springer-Verlag if and when these are accepted for publication.

The copyright covers the exclusive and unlimited rights to reproduce and distribute the article in any form of reproduction. It also covers translation rights for all languages and countries.

4. Manuscript must be written in English. One original and 3 sets of photocopies should be submitted to the following address.

Professor Toru Maruyama
Department of Economics
Keio University
2-15-45 Mita, Minato-ku, Tokyo
108-8345 Japan

5. **Offprints**: Offprints Order Form are to be sent you together with your page-proofs. Offprints will be available in lots of 100.

B. Preparation of Manuscript

1. Manuscripts must be double-spaced (not 1.5), with wide margins (at least 25 mm), and large type (at least 12 point) on one side of A4 paper. Any manuscript that does not meet these requirements will be returned to the author immediately for retyping.

2. All manuscripts would finally be composed using our Springer LaTeX macro package. If authors make manuscripts by word-processing software other than TeX, please follow our styles as possible. For authors who prepare their manuscripts by TeX, we strongly recommend to use our Springer TeX macro packages: cpmulti01 for plain-TeX and AMS-TeX, clmulti01 for LaTeX2.09. To make manuscripts to be double-sided, please use "referee" style-option (clmulti01) or "`\refereelayout`" command (cpmulti01). Above macro packages can be obtained from our ftp server (ftp.springer.de), in the directory: `/pub/tex`. For support, please contact `texhelp@springer.de`.

3. **The title page** must include: title; short (running) title of up to 60/85 characters; first names and surnames of all coauthors with superscript numerals indicating their affiliations; full street addresses of all affiliations; address to which proofs should be sent; fax number; and any footnotes referring to the title (indicated by asterisks*).

4. **Summary/abstract**: Each paper must be preceded by a summary/an abstract, which should not exceed 100 words.

5. **The Journal of Economic Literature index number (JEL classification)** should be indicated and the statement of the **1991 Mathematics Subject Classification (MSC) numbers** is desirable. You can check JEL classification with Internet at `http://ideas.uqam.ca/ideas/data/JEL`.

6. **Main text**: All tables and figures must be cited in the text and numbered consecutively with Arabic numerals according to the sequence in which they are cited. Please mark the desired position of tables and figures in the left margin of the manuscript.

Do not italicize, underscore, or use boldface for headings and subheadings.

Words that are to be italicized should be underscored in pencil.

Abbreviations must be spelled out at first mention in summary/abstract and main text. Abbreviations should also be spelled out at first mention in figure legends and table footnotes.

Short equations can be run in with the text. Equations that are displayed on a separate line should be numbered.

7. **References**: The list of references should be in alphabetical order and include the names and initials of all authors (see examples below). Whenever possible, please update all references to papers accepted for publication, preprints or technical reports, giving the exact name of the journal, as well as the volume, first and last page numbers and year, if the article has already been published or accepted for publication.

When styling the references, the following examples should be observed:

Journal article:
1. or [F-M] Freed, D.S:, Melrose, R.B.: A mod k index theorem. Invent. Math. **107**, 283-299 (1992)

Complete book:
2. or [C-S] Conway, J.H., Sloane, N.J.: Sphere packings, lattices, and groups (Grundlehren Math. Wiss. Bd. 290) Berlin Heidelberg New York: Springer 1988

Single contribution in a book:
3. or [B] Border, K.C.: Functional analytic tools for expected utility theory. In: Aliprantis, C.D. et al. (eds.): Positive operators, Riesz spaces and economics. Berlin Heidelberg New York: Springer 1991, pp. 69-88

8. **Citations in the text** should be either by numbers in square brackets, e.g. [1], referring to an alphabetically ordered and numbered list, or by the author's initials in square brackets, e.g. [F-M], or by author and year in parentheses, e.g. Freed and Melrose (1992). Any of these styles is acceptable if used consistently throughout the paper. In the third system, if a work with more than two authors is cited, only the first author's name plus "et al." need be given and if there is more than one reference by the same author or team of authors in the same year, then a, b, c should be added after the year both in the text and in the list of references.

9. **Tables** are to be numbered separately from the illustrations. Each table should have a brief and informative title. All abbreviations used in a table must be defined in a table footnote on first use, even if already defined in the text. In subsequent tables abbreviations need not be redefined. Individual footnotes should be indicated by superscript lowercase a, b, c, etc. Permission forms must be provided for any tables from previously published sources (same procedure as with figures; see below).

10. **Figures**: If you have access to suitable software, you can design your figures directly on a computer, but creation by other means is of course also possible. In any case the originals should be pasted into the camera- ready copy at the appropriate places and with the correct orientation. If necessary, the figures may be drawn overscale, but in this case suitably reduced copies should be pasted into the script.

If a figure has been published previously, acknowledge its source and submit written permission signed by author and publisher. The source must be included in the reference list. If a permission fee is required, it must be paid by the author. Responsibility for meeting this requirement lies entirely with the author.